Natural Products Chemistry

Sources, Separations, and Structures

Natural Products Chemistry

Sources, Separations, and Structures

Raymond Cooper

George Nicola

CRC Press
Taylor & Francis Group
Boca Raton London New York

CRC Press is an imprint of the
Taylor & Francis Group, an **informa** business

CRC Press
Taylor & Francis Group
6000 Broken Sound Parkway NW, Suite 300
Boca Raton, FL 33487-2742

© 2015 by Taylor & Francis Group, LLC
CRC Press is an imprint of Taylor & Francis Group, an Informa business

Printed on acid-free paper
Version Date: 20140520

International Standard Book Number-13: 978-1-4665-6761-0 (Paperback)

Library of Congress Cataloging-in-Publication Data

Cooper, Raymond, author.
    Natural products chemistry : sources, separations, and structures / Raymond Cooper, George Nicola.
       pages cm
    Summary: "Compounds isolated from nature have long been known to possess biological profiles and pharmaceutical potential far greater than anything made by man. However, they are notoriously cumbersome to isolate and challenging to synthesize, and the path of natural products to viable drugs is an arduous journey. This book presents a practical guide to gathering, isolating, and discovering new pharmaceuticals from nature. It emphasizes the challenges and advantages of products acquired from nature over traditional compounds such as those arising from combinatorial chemistry"-- Provided by publisher.
    Includes bibliographical references and index.
    ISBN 978-1-4665-6761-0 (paperback : acid-free paper) 1. Chromatographic analysis. 2. Biological products--Separation. 3. Microbial metabolites--Separation. I. Nicola, George (Chemist) II. Title.

RS189.5.C48C66 2014
615.1--dc23   2014016454

Visit the Taylor & Francis Web site at
http://www.taylorandfrancis.com

and the CRC Press Web site at
http://www.crcpress.com

Contents

SECTION I Basics

SECTION II   Selected Classes of Natural Products

SECTION III   Natural Product Contributions to Human Health

SECTION IV   *Nature's Pleasures and Dangers*

Foreword

This impressive natural products book is intended for undergraduates as an introduction to natural products for the first time, and for students studying pharmacognosy and related fields who wish to broaden their understanding of the field.

The book is not meant to be comprehensive, but rather an introductory text to natural products chemistry, and hence is broad rather than in-depth. It covers the major classes of natural product compounds and their sources. It is a preparatory text for more advanced studies on synthesis, biosynthesis, and mode of action studies. The authors have selected examples from marine life, plants, insects, and, importantly, actual examples taken from the pharmaceutical industry (e.g., anti-infective compounds), which are not always discussed or worked on in universities.

Natural products have played a central role in advancing synthetic and biosynthetic chemistry, medicine, and our understanding of nature.

The training of chemists and pharmacognocists in the area of microscale chromatographic purification and spectroscopy is ever increasing to tackle the challenging questions in bioorganic chemistry and molecular biology.

Over the lifetime of this author, dramatic changes have taken place in our approaches for the isolation, structure determination, and mode of action studies of chemicals from nature: from isolation of large quantities to even nanogram quantities.

It continues to be essential to properly train natural products chemists in the area of the analytical challenges in the field of botanical medicines (supplements), and to provide confidence to encounter ever-increasing new challenges to solve critical biological questions at the interface of bioorganic chemistry and molecular biology.

The book is timely as there is currently a gap in the formal education students are taught in chemistry-related courses pertaining to natural products. We are also in an era when students are increasingly conscious of their environment, and of the value in becoming further aware and preserving our natural world.

From our recent renewed interests and passion of marine and terrestrial ecosystems, this book serves as a most wonderful segue into chemistry and will hopefully pique interest to pursue further studies in natural products. The authors, with an almost unprecedented broad experience in various research subjects and countries, are uniquely suited to handle such a book.

Koji Nakanishi
Centennial Professor of Chemistry (Emeritus)
Columbia University
New York, New York

Prologue

Where have all the flowers gone? Long time passing...

As the 1960s folk singers were strumming their guitars, natural products research was reaching its zenith. The research on complex natural molecules from microbial, terrestrial, and marine sources continued unabated. The pharmaceutical industry in particular had a "love fest" with natural products structures and sought chemical leads targeted as anti-infectives and antifungals.

There are some wonderful modern success stories. For example, Africa's gift to the world, the periwinkle *Catharanthus rosea* and the discovery of vincristine, were developed to treat childhood leukemia. Then there is the anticancer drug, taxatere, also known as paclitaxel. This complex natural product, isolated from the Pacific yew for all its synthetic complexity, became a successful commercial drug due to a combination of natural products ingenuity and medicinal chemistry; obtaining the precursor to paclitaxel through fermentation and the final form developed with a semisynthetic pathway. This example is only one of a long list of remarkable modern discoveries.

However, as researchers in these drug discovery programs soon realized, with the screening emphasis moving away from antimicrobial and antifungal assays toward receptor-based assays, coupled with molecular biology, the number of natural products leads suitable for drug development began to dry up. The research focus and direction changed from searching for leads from nature to moving instead toward *in silico* design and combinatorial chemistry. Natural products lead discovery programs in the pharmaceutical industry were disbanded. In hindsight, the drug industry was perhaps too hasty to hitch its wagon to combinatorial chemistry and one can question that wisdom. Furthermore, particularly in the United States, we are not funding at the same levels of academic training of researchers to continue this work. Thus, the talent and expertise, previously available to uncover remarkable structures and mine the treasures of natural products isolation and structure determination from complex mixtures to complex structures, alas, is disappearing.

Do not be alarmed! First, natural products will never entirely go away. Second, the training and knowledge required to be a successful natural products researcher can be applied to many other scientific disciplines. The technical capabilities of a natural products chemist working with "unknowns" and learning how to isolate minute quantities of pure material from complex mixtures can be transferred to other science disciplines and these are outlined in Figure P.1. The figure shows how many key research areas are amenable to trained natural products scientists, and their technical capabilities are well suited to contribute and to impact these various disciplines. There is a lot of interest, not only from the pharmaceutical industry on downstream drug development applications, but the industries related to botanical and herbal remedies, supplements, and bioalgae are expanding too. Also, the

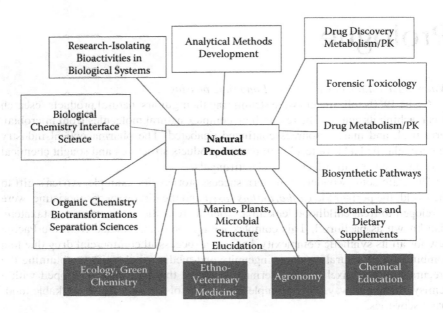

FIGURE P.1 Examples of various scientific disciplines where the training and expertise in natural products chemistry can be successfully applied.

technological steps applicable to the separation of natural products span aspects of analytical sciences and chemistry. Examples include analytical, quality assurance (QA), metabolism studies, and pharmacokinetic (PK) research applications used in today's chemical, agrochemical, and pharmaceutical industries.

Where have all the chemists gone? Long time passing...

Acknowledgments

The authors thank the following for all their help: Professor Lars Bohlin, Uppsala University, Sweden, for kind comments on the Prologue; Dr. Vivien Shanson, formerly at Imperial College, London University, United Kingdom, for her insights; and Hilary Rowe and Laurie Schlags at Taylor & Francis/CRC Press for their patience and kind support.

The authors also thank Steven Foster at Steven Foster Group Inc., Eureka Springs, Arkansas, for permission to use the photographs of the plants and to singer Pete Seeger for his inspiring words.

And finally, one of us (RC) wishes to thank the library staff and the colleagues at the Missouri Botanical Gardens in St. Louis, for all their support and advice.

Authors

Raymond Cooper, Ph.D., was born in the United Kingdom, received his Ph.D. in organic chemistry and spent 15 years in discovery and research positions in the pharmaceutical and biotech industries. He then moved to the dietary supplements industry, first with Pharmanex (NuSkin). There, he researched and developed novel Traditional Chinese Medicines including Cholestin, Cordymax, and Te-Green. Dr. Cooper continued in the nutrition industry to create and develop several more innovative botanical solutions and products. Currently, he is a visiting professor and lecturer at the Hong Kong Polytechnic University providing courses on natural products chemistry, dietary supplements, and regulatory and patent strategies. He is also a science writer and editor of the *Journal of Alternative and Complementary Medicine*; a fellow of the Royal Chemical Society, United Kingdom; and visiting professor at the School of Pharmacy, London University, United Kingdom. He has edited five books, most recently *Botanical Medicine: from Bench to Bedside* (Liebert, 2009), and published more than 100 peer-reviewed scientific articles.

George Nicola, Ph.D., MBA, grew up in New Haven, Connecticut, and attended the University of Massachusetts Amherst, graduating in 2000 with a BS in biomedical computing. He went to graduate school at the Medical University of South Carolina in Charleston where he earned his Ph.D. in biochemistry in 2005. He has been a postdoctoral fellow at The Scripps Research Institute and was awarded a Research Scholar Fellowship from the American Cancer Society in 2007. Dr. Nicola has also held an appointment as adjunct professor of chemistry at the University of San Diego in 2008. In 2009, he earned an MBA from the Rady School of Management, University of California, San Diego (UCSD). He has also held an appointment as project scientist at the Skaggs School of Pharmacy and Pharmaceutical Sciences at UCSD.

Introduction

OBJECTIVES OF THIS BOOK

Compounds isolated from nature have long been known to possess biological profiles and pharmaceutical potential far greater than anything made by man; however, natural products are notoriously cumbersome to isolate and very challenging to synthesize. For example, the stringent requirements and resource allocations of the pharmaceutical industry make the path of natural products to viable drugs an arduous journey.

This book is intended for individuals who are either taking classes in natural products, or are interested in the field of natural products structures and their pharmaceutical applications. As an introductory guide, this text is intended to help the next generation of motivated entrepreneurial students seeking practical wisdom on extract fractionation, compound isolation, identification of pharmacological properties, and structure elucidation.

Whereas many academic texts covering chemistry and biology focus on fundamental science that will be used primarily in research settings, this text broadens the scope of applications and draws upon examples from various sources. Natural products chemistry in the United States, as taught today, draws its examples mainly from marine chemistry or plant chemistry, however, there is a fascinating and rich world of fermented (microbial and algal) products leading to complex structures. Therefore, the book draws upon examples from the microbial world too. This is a source of bioactive metabolites, not traditionally accessible in academia: more the mainstay of the pharmaceutical industry. Second and perhaps more importantly from the point of view of secondary metabolism and sourcing, one of the interesting areas being pursued in marine chemistry is the symbiosis between the host organism (e.g., mollusk, coral), which may not by itself produce the bioactive molecules, but act as the template for microbes to grow and yield diverse structures.

Many natural products are extracted from plant, marine, and microbial sources using organic solvents. The book provides examples from published literature on isolation processes and large-scale production. Again, these examples are drawn mainly from the pharmaceutical industry providing many examples of wonderful microbial products, produced in fermentation media and subsequently extracted using a variety of techniques. Many microbial products are water soluble and remain in the aqueous phase (e.g., peptides, alkaloids, glycosides, etc.). The pharmaceutical industry relies on these isolation techniques, but graduates fresh out of school are not always trained or exposed to these challenges. Regarding the isolation of polar molecules, it is important to emphasize such chromatographic techniques as ion exchange chromatography and countercurrent techniques, as well as the more classical organic-phase separations. Thus, the book is intended to be more applied-based than many other natural products books, drawing upon the sources, separations, and structures of various natural products.

1

The book is intended as an introductory text to the vast world of natural products chemistry. This will be a much more readable text, with easy-to-grasp concepts that are directly applicable to undergraduate-level courses in chemistry and biology. With a strong emphasis on the practicality of natural products chemistry to drug discovery, it will be especially appealing to those students eager to learn real-world applications in their classroom instruction. Examples are chosen that give students a breakdown of the cumbersome process of collecting a compound from nature, isolating the active ingredient, and taking it all the way through the steps required to prepare it into a pure compound and in some cases into lifesaving medicine—all with concise and easy-to-understand terms.

Select topics and natural products structures will be chosen to illustrate the fundamentals and review several major classes of compounds (i.e., steroids, alkaloids, terpenes, etc.) before embarking on specific structures. The book is not intended as a comprehensive treatise on the plethora of natural products. Excellent texts are available. Nor is it a basic chemistry text, as we assume the reader has at least a rudimentary understanding of organic chemistry, chemical structures, and stereochemistry. We only offer a general commentary on the theory of chromatographic separations and on the background to spectroscopic techniques. Our quest is to show how we apply these tools to achieve the desired goal of isolation and structure determination of a target natural product.

This book is written in a style that reflects natural products research: multifaceted, with contributions from botany, chromatography, spectroscopy, and pharmacology to achieve at least one purpose: to identify and elucidate the structure of a natural product from its source. Further, techniques of isolating and determining the structures of natural products are provided. Main themes include chromatographic separations and spectroscopic ways to elucidate structures (^{1}H-NMR, ^{13}C-NMR, mass spectrometry, and X-ray analysis; for more information, see the "Glossary" at the end of the book). Examples are drawn from Western medicinal and pharmaceutical research, botanical and drug research, and from other more traditional sources, such as Traditional Chinese Medicine (TCM) and herbal medicines.

We present the material in the form of examples and case studies, offering the background chemistry, chromatographic routes, and structural features to make as complete and (we hope) as interesting a story as possible. The techniques are intertwined with core material defining primary and secondary metabolites, the isoprene routes, and other key metabolic pathways.

Several of the important classes of natural products will be covered in the text and examples are provided to illustrate their sources and structural features. This is not a comprehensive list and we cannot do justice to treatises that came before us. We have chosen examples that we think are important and illustrate key aspects: these include the classes of carbohydrates, sugars, lipids, fatty acids, omega oils, prostaglandins, terpenes, steroids, and nitrogen-containing molecules (e.g., proteins, peptides, amino acids, nucleic acids, alkaloids). This section is followed by a chapter on specialty topics. We have chosen examples from the rich diversity of various natural sources, including toxins in nature; the anticancer drug, taxatere; the antimalarial drugs, quinine and artemisinin; popular alternative natural botanicals sold as supplements, for example, green tea catechins, red yeast rice, and ginkgo; alkaloids; natural sweeteners; selected vitamins; carotenoids; and β-lactam antibiotics.

Natural products need to be related as stories—from their source to their use. Natural products research is a fascinating branch of chemistry and we hope you will feel our enthusiasm for the subject through the selection of topics herein.

THE IMPORTANCE AND ROLE OF NATURAL PRODUCTS

It is perhaps of no surprise that mankind sought chemicals from nature. We realized that plants were both foods and medicines. Plants were available, easily collected, and produced seeds for future harvests. Early civilizations learned of planting and planned farming, leading today to well established food and tobacco industries. Major commercial crops include wheat, rice, corn, tobacco, sugarcane, tea, and coffee. Today, there are many examples of important natural products as drugs and some are reviewed in the text.

As source material to begin an investigation, plants were abundant and easily obtained. More and more chemicals were isolated and their chemical structures elucidated. Plants have been used for medicinal purposes across history and cultures and even across species. Indeed, a majority of the world still relies heavily on natural products as herbal remedies for their primary health care. In fact, over the past several decades, literature, both scientific and popular, reflects an increased interest

HISTORICAL NOTE

The chemicals, morphine and strychnine, were isolated in 1815 and 1819, respectively. However, their actual chemical structures were not elucidated for another 100 years. Research during the 19th century centered on pigments from flowers, which were used as colored dyes and resulted in their commercialization. Chemicals from plants and from underground oils spawned large industrial chemical entities. Then in the 20th century, numerous examples of pharmaceutical and over-the-counter (OTC) drugs were derived directly from compounds found in nature or indirectly through chemical modification of the basic structure. As the chemical techniques and chemical knowledge became more refined and sophisticated, we were able to take even more advantage of searching for natural products from marine and microbial sources.

The emphasis of seeking medicines from plants shifted after World War II to using microbes, and harnessing the improvements in fermentation technology led to the discovery of many new microbial products. For example, once the fermentation technology was firmly established, large amounts of penicillin and many other lifesaving anti-infectives from microbial sources were obtained.

in natural products by the general public and has helped fuel a greater scientific awareness of both drugs and botanical medicines from nature.

With the increasing movement of people across countries, there is an accompanying movement of their respective traditional medicines. Although globalization of medicine has typically been thought of as being the movement of Western medicine to developing countries, there is now, in China, for example, significant discussion of the "globalization of Traditional Chinese Medicine" and discussion of how to improve the quality of Chinese medicine products and interest in developing a body of research evidence for their effectiveness. This is particularly relevant given the establishment of Traditional Chinese Medicine programs and practitioners around the world. The interest in traditional herbal medicines has also contributed to a resurgence of interest in Western herbal medicine, particularly in the United States and Europe (phytomedicines), and a desire for information about the sourcing, safety, and efficacy of medicines. This study of chemicals from nature as sources of drugs is called *pharmacognosy*.

THE WORLD OF PHARMACOGNOSY

Pharmacognosy, commonly referred to as the pharmacology of natural products, derives from two Greek words, *pharmakon*, or "drug," and *gnosis*, or "knowledge." Its scope includes the study of the physical, chemical, biochemical, and biological properties of drugs, drug substances, or potential drugs or drug substances of natural origin as well as the search for new drugs from natural sources. Research in pharmacognosy often embraces areas of study of phytochemistry, microbial chemistry, biosynthesis, biotransformation, chemotaxonomy, and other biological and chemical sciences.

The methods used in the extraction of plant material can influence the chemical composition of the resulting extracts and potentially the biological activity. The more information on the product used that is provided in research publications, the greater will be the ability to compare among studies and understand differences in results that may emerge. Using many examples of natural products from a variety of sources, this book addresses the characterization of natural products, the techniques used to successfully isolate the chemical components, and techniques used to elucidate their structures.

There are challenges in conducting sound scientific studies of botanicals: from the sourcing of appropriate products to details on the preparation of the products and understanding the types of scientific inquiry that will advance this field. This discussion also includes challenges to the dietary supplement industry (particularly in the United States) to continue to improve the quality of botanical products and to increase the industry's commitment to facilitating the collection of data that will provide the evidence base for the safety and effectiveness of the products being produced. This will reassure current consumers and health care providers and bring into the user groups clinicians and a wider public who all seek treatments that are safe and effective and that have the fewest side effects possible.

The role of pharmacognosy in furthering the quality of botanical preparations and the quality of clinical research is evident throughout. Examples are provided of high-quality scientific research required to achieve higher-quality preclinical and clinical studies of herbal preparations and better quality herbal products. Topics in the book were chosen to illustrate many of the issues and a variety of scientific approaches that address the challenges in botanical research.

FURTHER READING

D. Barton, K. Nakanishi, and O. Meth-Cohn. 1999. *Comprehensive Natural Products Chemistry.* Elsevier.

G. Brahmachari. 2012. *Bioactive Natural Products: Opportunities and Challenges in Medicinal Chemistry.* World Scientific.

S. M. Colegate and R. J. Molyneux. 2007. *Bioactive Natural Products: Detection, Isolation, and Structural Determination.* 2nd ed. Taylor & Francis.

P. M. Dewick. 2009. *Medicinal Natural Products: A Biosynthetic Approach.* 3rd ed. Wiley.

X. T. Liang and W. S. Fang. 2006. *Medicinal Chemistry of Bioactive Natural Products.* Wiley.

L. Mander and H.-W. Liu, eds. 2010. *Comprehensive Natural Products II: Chemistry and Biology.* Elsevier Science.

S. P. Stanforth. 2006. *Natural Product Chemistry at a Glance.* Wiley-Blackwell.

V. E. Tyler, L. R. Brady, and J. E. Robbers. 1988. *Pharmacognosy.* 9th ed. Lea & Febiger.

G. H. Wagman, 1989. *Natural Products Isolation: Separation Methods for Antimicrobials, Antivirals & Enzyme Inhibitors.* Elsevier Science Limited.

Section I

Basics

Section I

Basic

1 Natural Products Sources

Sand is a natural product. But we will not cover the chemistry of silica. Natural products of interest are those organic compounds that contain at a minimum the element carbon. Carbon (chemical symbol: C) has a valency of 4 and is bound to another carbon either as a single, double, or triple bond. Carbon is also attached to other elements, typically other carbons, hydrogen (H), oxygen (O), and nitrogen (N); and also sulfur (S) and phosphorous (P). The various combinations lead to the bewildering array of natural products structures: some simple and others complex. Examples are provided throughout the book.

1.1 PRIMARY AND SECONDARY METABOLITES

Metabolites are intermediates in metabolic processes in nature and are usually small molecules. A primary metabolite is directly involved in normal growth, development, and reproduction, for example, fermentation products (ethanol, acetic acid, citric and lactic acid) and cell constituents (lipids, vitamins, and polysaccharides). A secondary metabolite is not directly involved in those processes and usually has a function but is not that important for the organism (e.g., antibiotics, pigments, and carotenoids).

Metabolic processes and enzymatic reactions begin from simple building blocks, which are outlined in Figure 1.1. The building blocks and pathways are further elaborated upon in Chapter 4. These metabolites are divided into various classes depending on the functionality of the molecule (e.g., phenolic, alkaloid, steroid), and are discussed in detail in Chapters 6 and 7. Several simple but important natural products structures and their sources in nature are described next.

EXAMPLE 1.1: PHENOLIC COMPOUNDS FROM THE WILLOW BARK

The active ingredient in willow bark is salicin, converted in the body into salicylic acid (Figure 1.2). The presence of a phenol is recognized by the attachment of an OH group to the benzene ring. Substitutions on the phenolic group of various natural products include, but are not limited to, methyl, acetyl, and ether linkages.

HISTORICAL NOTE

Chewing the willow bark from the tree *Salix alba* reduced fever and inflammation, and was recognized by ancient Greek physician Hippocrates to possess health benefits. In the 1800s, pharmacists created salicylic acid in its acetylated form (acetylsalicylic acid), more commonly known as aspirin. Aspirin was first isolated and synthesized by Felix Hoffmann, a chemist with the German company Bayer and marketed in 1897. Thus, the most widely used drug in the world continues to be aspirin.

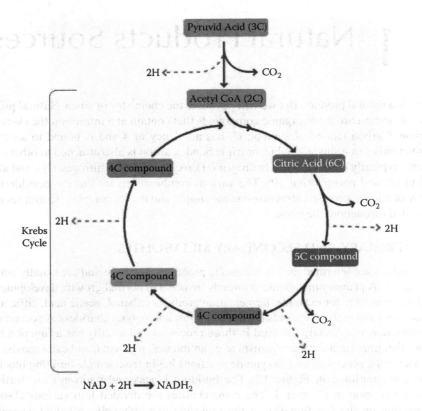

FIGURE 1.1   The Krebs cycle indicating that metabolic processes and enzymatic reactions begin from simple building blocks.

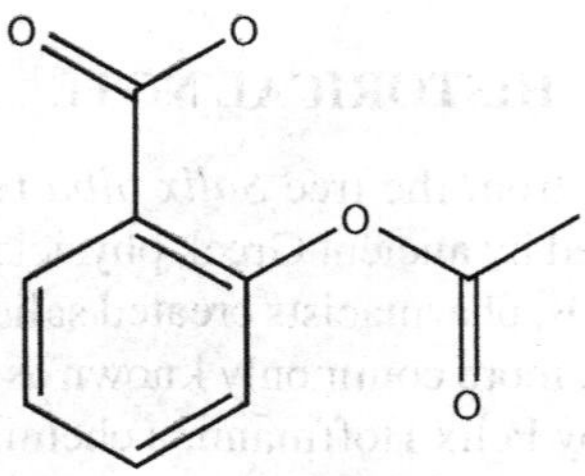

FIGURE 1.2   The active ingredient in willow bark is salicin, which is converted in the body into salicylic acid. Salicylic acid is more commonly known as aspirin.

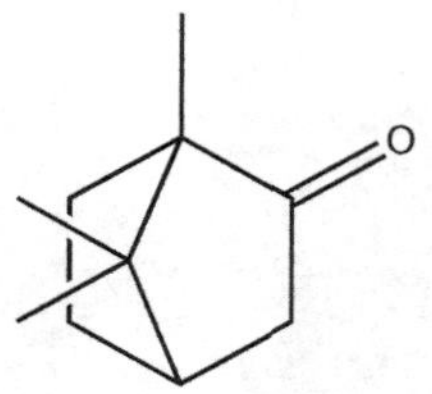

FIGURE 1.3   Camphor is a terpenoid found in the wood of the camphor laurel, a large evergreen tree found in parts of Sumatra and Borneo.

EXAMPLE 1.2: THE TERPENE CAMPHOR

Camphor (Figure 1.3) is a terpenoid with the chemical formula $C_{10}H_{16}O$. It is found in the wood of the camphor laurel (*Cinnamomum camphora*), a large evergreen tree found in parts of Sumatra and Borneo. It is a volatile oil and is often used for its scent.

EXAMPLE 1.3: ALKALOIDS AND THE PRESENCE OF NITROGEN

Alkaloids form a very important class of natural products, widely dispersed in nature with a variety of simple and complex structures. Nitrogen is present in alkaloids as part of cyclic ring structures. One very well-known alkaloid is nicotine, found in tobacco (Figure 1.4). The compound contains two nitrogens, one in the pyridine ring and the second as a reduced (tetrahydro)–pyrrole ring.

EXAMPLE 1.4: A STEROID FROM THE GARDEN

In Western medicine, an early example is the steroid molecule known as digitalis (Figure 1.5) from the garden variety of foxglove (Latin name: *Digitalis purpurea*) (Figure 1.6). Foxglove, including the roots and seeds, contains several deadly physiological and chemically related cardiac and steroidal glycosides. Digitalis was used for the first time in 1785, and thus began the era of modern therapeutics. Digitalis is very toxic and must be very carefully administered. Digitalis is an example of a drug derived from a plant used by folklorists and herbalists but it is difficult to determine the amount of active drug in herbal preparations. It can still be prescribed for patients in atrial fibrillation, especially if they have congestive heart failure. However, an overdose of digitalis can cause anorexia, nausea, vomiting, and diarrhea.

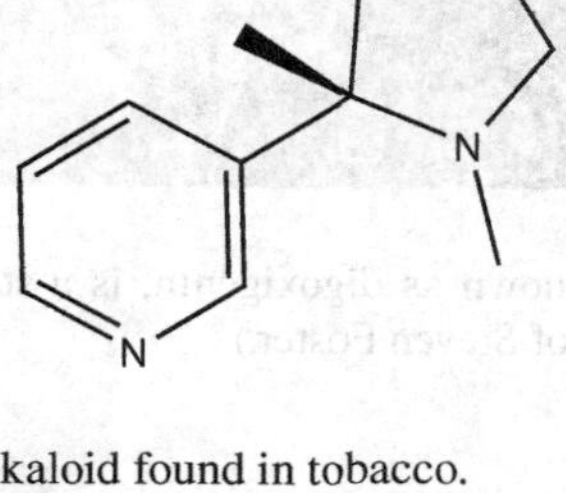

FIGURE 1.4   Nicotine is an alkaloid found in tobacco.

FIGURE 1.5 Digitalis, also known as digoxigenin, is a steroid molecule extracted from foxglove. (Photograph courtesy of Steven Foster.)

FIGURE 1.6   Foxglove (Latin name: *Digitalis purpurea*) contains several deadly cardiac and steroidal glycosides. (Courtesy of Steven Foster.)

EXAMPLE 1.5: SULFUR IN NATURE

Garlic, known as *Allium sativum*, is a species in the onion genus *Allium*. Garlic has a history of use for over 7000 years and is used for both culinary and medicinal purposes. There are a number of organosulfur compounds present in garlic and shown in Figure 1.7.

1.2   ETHNOBOTANY AND TRADITIONAL SOURCES OF NATURAL PRODUCTS

Historically, mankind has been intrigued by the power and potential of plants in nature. Ancient texts attest to the knowledge passed down through generations on the beneficial effects of plants. Wisdom was gained on how to extract ingredients as foods, medicine, and mood enhancers long before we knew how the chemicals themselves worked. At the time of the Renaissance and through the Reformation period, the tools were found to explore the chemistry of the natural ingredients. Today, we embrace the science of ethnobotany, which is the scientific study of the relationships that exist between people and plants. The discipline requires a variety of skills: botany (the identification and preservation of plant specimens); anthropology, to

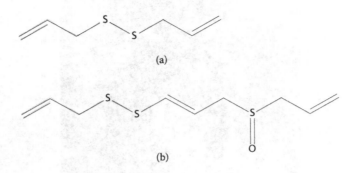

(a)

(b)

FIGURE 1.7 Garlic, a species in the onion genus, has a history of use of over 7000 years for both culinary and medicinal purposes, and contains sulfur compounds. Above: (a) Diallyldisulphide and (b) ajoene. (Photograph courtesy of Steven Foster.)

understand the cultural concepts around the perception of plants; and natural products chemistry.

As sources of health products, the majority of the world continues to heavily rely on herbal remedies for their primary health care. Ancient civilizations of both China and India have provided a wealth of knowledge on the use of traditional medicines.

Traditional Chinese Medicine (TCM) has been used for thousands of years. The basis and philosophy of TCM centering on the *qi* is not well understood in Western practice. Many TCM formulas are mixtures of herbs and plants, used in combination and prepared as teas. There is also the use of various medicinal mushrooms. One example is the mushroom *Ganoderma lucidum*, known also as Reishi and Ling Zhi, which has been used in TCM for thousands of years as an herbal tonic (see also Section 5.8).

Ayurvedic medicine is used in India. Ayurveda stresses the use of plant-based medicines and treatments. Hundreds of plant-based medicines are employed, including two commonly used spices: cardamom and cinnamon.

Kampo medicine is used in Japan and much of its origins derive from TCM formulas. They also highly respect their medicinal mushrooms. An example is the *Agaricus blaze* mushroom, a highly popular medicine that is used by close to 500,000 people, particularly by cancer patients. The second most used medicine is derived from another mushroom, an isolate from the shiitake mushroom (*Lentinula edodes*).

Africa is endowed with many traditional healing plants that can be used for medicinal purposes. Out of approximately 6400 plant species used in tropical Africa, more than 4000 are used as medicinal plants. Examples include pygeum (*Prunus africana*) to increase the ease of urination and reduce inflammation and cholesterol deposits. The bark is made into tea. In traditional African practice, *Momordica balsamina* is known to lower blood glucose levels.

Native Americans also have a long history of use of traditional medicines. They use sage, which was believed to heal multiple problems of the digestive system. Remedies for common colds include American ginseng; herbs for aches and pains include Pennyroyal and hops; and dogwood and willow bark are remedies for fever.

1.3 SOURCING OF NATURAL PRODUCTS

With respect to the sourcing of natural products from plants, first, attention needs to be paid to the correct identification of the species. Second, due to variability in the growing conditions, it is important to keep in mind that the same identified species found in different geographical locations might yield different metabolites or ratios of metabolites. Finally, the selection of the plant part is very important: flowers, fruiting bodies, berries, bark, seeds, or stems. Metabolites vary within the whole plant as fresh or dried material, or from the roots or the aerial parts. Thus, the botanical extracts are complex mixtures of multiple compounds. For chemical analysis, reference standards may be difficult to generate or obtain. Also, the stability of chemicals in the extracts may sometimes be difficult to control. Some advances include:

1. Use of Global Positioning System (GPS) site monitoring, particularly in the case of plants found in the wild.
2. More in-depth taxonomy and chemical ecology to compare species grown under different environmental conditions.
3. More chemical analyses and testing of the materials to ensure correct identification.

1.4 SOURCES OF MICROBES

Bacteria and fungi have been an invaluable source for discovering antimicrobial agents. Bacteria and fungi are often found growing in soils and on rotting vegetation, and are symbiotic with marine animals.

The techniques in microbiology to enhance strain development and fermentation yields, followed by improved downstream purifications, have led to some remarkable discoveries, penicillins, cephalosporins, and tetracyclines, to name but a few antimicrobials. They are presented in more detail in Chapter 9.

1.5 MARINE SOURCES

Marine collections, in contrast to plants, offer different challenges. The challenge is sourcing the material in oceans and deep waters and preserving the ecology, particularly the fragile coral reefs. The chemistry of marine natural products will be influenced by different variables, for example, currents and sediments, pH levels, atmospheric constituents, metamorphic activity, and ecology.

However, corals, sponges, fish, and other marine microorganisms have provided a wealth of biologically potent chemicals with interesting inflammatory, antiviral, and anticancer activity. For example, curacin A is obtained from a marine cyanobacterium and shows potent antitumor activity. Other antitumor agents derived from marine sources include the bryostatins, described in Chapter 10, and marine toxins are presented in Chapter 14.

1.6 ANIMAL SOURCES

Interestingly, animals have yielded new natural chemical entities. For example, a series of antibiotic peptides was extracted from the skin of the African clawed frog and a potent analgesic compound called epibatidine (Chapter 14) was obtained from the poisonous skin extracts of the Ecuadorian frog.

1.7 VENOMS AND TOXINS

Even more remarkably, we have discovered venoms and toxins not only from plants and microbes but snakes, spiders, scorpions, and insects. Generally these compounds are extremely potent because they often have very specific interactions with a macromolecular target in the body. As a result, they have provided researchers with important tools in studying receptors, ion channels, and enzymes.

Many of these toxins are polypeptides such as bungarotoxin from cobras. However, nonpeptide toxins such as tetrodotoxin from the puffer fish (Chapter 14) are also extremely potent.

SUMMARY

In this chapter we highlight the difference between primary and secondary metabolites, provide the definition of ethnobotany, and discuss traditional sources of natural products. The sourcing of natural products from terrestrial, marine, microbial, animal, and insect species is indicated.

QUESTIONS

1. Provide an example of a natural product containing one of the following: (a) a carbon bound as a single bond, (b) a compound with a carbon double bond, and (c) a compound with a carbon triple bond.
2. Name five cultures still using natural products as traditional medicines.
3. Give an example of at least one plant used by each culture named in Question 2.
4. Why do indigenous populations mainly use plants?
5. List some of the challenges a researcher faces in seeking natural products material.
6. What is organic?
7. What elements are typically in natural products?
8. Where do you find natural products in nature?
9. What is the study of natural products called?

2 Extraction and Separation of Natural Products

Naturally occurring chemical compounds are found in a variety of sources and invariably are present as mixtures. It is the task of the natural products chemist to isolate and purify specific components, usually to homogeneity, in order to identify the chemical structure. A variety of separation techniques is available. Generally, the first step is the use of a solvent (aqueous, organic, or liquid CO_2, etc.) to dissolve organic material, and the solute is separated from the marc (plant residue) or the mycelia (fermentation). By concentrating the solvent an ideal outcome would be crystallization! This is rare and generally other separation steps are required.

2.1 WATER-STEAM DISTILLATION

In its simplest form, water is boiled and the water-steam vapors are passed over and through the source material, releasing and carrying the (volatile) chemicals away from the residue. This technique is commonly used to extract volatile plant oils as shown in Figure 2.1. In many cases, as the steam condenses and cools, the oils float to the top and are easily removed from the aqueous phase.

2.2 SUPERCRITICAL FLUID EXTRACTION

Supercritical fluid extraction (SFE) is used for separating one component from the matrix with the application of supercritical fluids as the extracting solvent. Carbon dioxide (CO_2) is the most used supercritical fluid. Extraction is usually from a solid matrix, but can also be from liquids. Extraction conditions for supercritical carbon dioxide require a temperature around 31°C and a critical pressure of 74 bar.

2.3 SOLVENT PARTITIONING

Upon extraction of the solids and release of desired organics into the extraction solvent, the most common next step is a liquid–liquid extraction, taking advantage of mixing two (or sometimes three) immiscible solvents, for example, water and ether.

The common rule of thumb is that polar compounds go into polar solvents (e.g., amino acids, sugars, and proteins remain in water), whereas the nonpolar compounds generally remain in the organic phase (e.g., steroids, terpenes, waxes, and carotenoids are normally extracted into a solvent such as ethyl acetate).

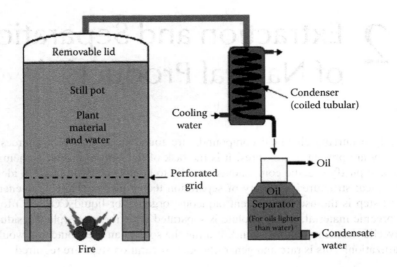

FIGURE 2.1 Simplified drawing of a steam distillation apparatus. When steam condenses and cools, the oils then float to the top and separate from the aqueous phase.

Extract Flow to Isolate Natural Products from Their Source (e.g., Marine and Plant Products)

1. Source the natural material (fresh or in dried form)
2. Grind material to powder
3. Extract with an organic solvent (liquid extraction)

4. Filter, remove *marc*
5. Retain the eluate

6. Separate between two immiscible solvents (solvent partition)

Retain organic layer and concentrate.

Note: In the case of a microbial source, upon fermentation the whole broth is either extracted in its entirety with an organic solvent, followed by filtration. Alternatively, the whole broth is filtered (centrifuged) to remove the cell mass and the mycelia, then the filtrate is partitioned with an immiscible solvent or adsorbed onto a solid phase resin.

2.4 REFINED ISOLATION TECHNIQUES AND CHROMATOGRAPHY

2.4.1 SEPARATIONS OF NONPOLAR COMPOUNDS

There are three common separation techniques. Thin layer chromatography (TLC) is by far the simplest, cheapest, and most rapid. Mixtures are applied to precoated silica gel plates, which are then immersed in a chamber containing a sufficient amount of the eluting solvent, to allow for solvent migration. As the solvent migrates up the plate the mixture is separated on the coated plate due to each component having a different retention time. TLC is used mostly for small quantities of material under analysis and is a very versatile technique for nonpolar compounds. For separation of larger quantities of nonpolar materials, silica gel column chromatography (often referred to as normal phase) is commonly used. In this case, an eluent is passed through a column packed with silica gel. The mixtures of compounds are separated and are eluted off the silica gel column in the eluting solvent. Two methods are generally used to prepare a column: the dry method and the wet method. For the former, the column is first filled with dry stationary phase powder, followed by the addition of the mobile phase, which is flushed through the column. In the latter case, a slurry comprised of eluent and the stationary phase powder is mixed and poured into the column. The stationary phase is often silica gel, but alumina, cellulose, and polyamide are quite commonly used as well.

The mobile phase (eluent) is either a single solvent or more commonly a mixture of miscible solvents. The eluent can be optimized in small-scale pretests, often using TLC with the same stationary phase. One popular form of normal phase chromatography is called "flash chromatography" based on the same principle as described earlier, where the particle size of silica gel is much smaller, typically 250–400 mu, and the mobile phase is passed rapidly over the stationary phase using medium pressure.

For more volatile and nonpolar compounds, gas chromatography (GC) is used. In this case, a carrier gas (e.g., nitrogen/helium) is passed over a coated and heated column. This method has been successfully used for the separation of oils, volatiles, terpenes, and esters of fatty acids.

An example is the separation of long chain fatty acids in fish oils. In order to achieve a well-resolved GC analysis, it is often necessary to convert these fatty acids to their respective methyl esters. A polar stationary phase may be used for the chromatographic separation and a typical analysis of the methyl ester derivatives of cod liver oil is shown in Figure 2.2 using a fused silica column coated with Carbowax 20M™, a packing material that contains polyethylene glycol (PEG) polymers bound to silica.

2.4.2 SEPARATIONS OF POLAR COMPOUNDS

There are several different chromatographic systems amenable to the separation of polar compounds. Polar compounds are usually soluble in aqueous systems and if the compound of interest possesses a charge (due to the presence of either positive or negative functional groups) then ion exchange chromatography is very useful. The

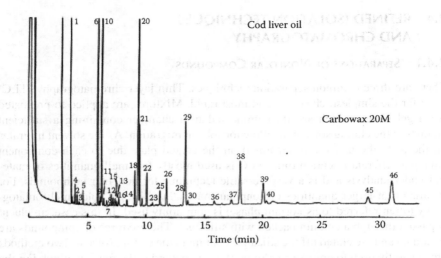

FIGURE 2.2 Gas chromatography (GC): Analysis of fatty acids (in the form of methyl esters) on a fused silica column coated with Carbowax 20M.

resin is often made from polymeric material, coated or bound with a resin. There are four main types of ion exchange resin differing in their functional group:

1. Strongly acidic (typically, sulfonic acid groups, e.g., sodium polystyrene sulfonate)
2. Strongly basic (quaternary amino groups, e.g., trimethylammonium groups)
3. Weakly acidic (mostly carboxylic acid groups)
4. Weakly basic (primary, secondary, and ternary amino groups, e.g., polyethylamine)

Mechanistically, for cation exchange chromatography, the resin retains the positively charged cations because the stationary phase displays a negatively charged functional group, shown in the following equation:

$$R - X^-C^+ + M^+B^- \rightleftarrows R - X^-M^+ + C^+ + B^-$$

With respect to anion exchange chromatography, anion resins retain the anions due to the presence of the positively charged functional groups:

$$R - X^+A^- + M^+B^- \rightleftarrows R - X^+B^- + M^+ + A^-$$

In either case, the ionic strength of either C^+ or A^- in the mobile phase can be adjusted to shift the equilibrium position and thus the elution time.

2.5 CHARCOAL

Black, refined and of fine mesh or macroporous, charcoal is very abundant. There are several advantages to using charcoal: very cheap, plentiful, and can be used with either aqueous or organic solvents or a combination of both (e.g., aqueous methanol).

The exact mechanism by which charcoal interacts with solutes is not well understood, however, it is often used to "decolorize" and remove impurities.

Flow Diagram Showing the Efficient Separation of a Small Peptide Using Ion Exchange and Charcoal Chromatography from a Fermentation Broth

Fermentation broth (110 liters, adjusted to pH 7)

 Filter

Filtrate

 Adsorb onto BioRad AG 50X8 (H⁺), 2 liters
Elute with 0.5 n NH₄OH

Eluate (78.8 g)

Adsorb onto BioRad AG1X8 (HCO₃⁻)
Elute with CO_2 – saturated water (carbonic acid ~ pH 4.5)

Eluate (3.4 g)

 Lyophilize

Chromatograph on charcoal, 1 liter
Elute gradient of 0% to 20% aqueous MeOH

1.3 g pure compound
Sch37137 (Figure 2.3)

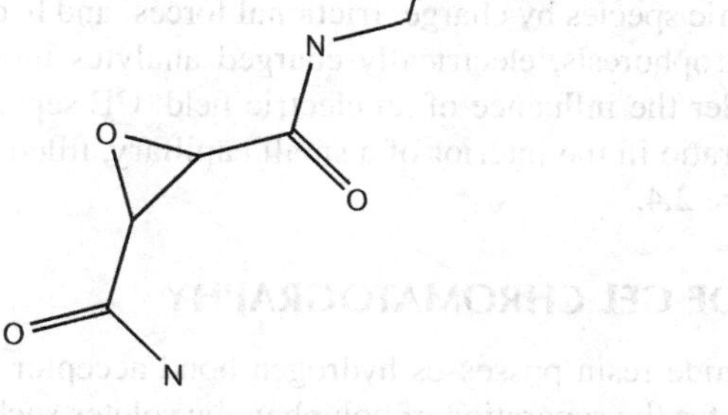

FIGURE 2.3 Structure of the antifungal compound, SCH37137. (From R. Cooper et al., 1988, *J Antibiot (Tokyo)* 41(1):13–19.)

2.6 REVERSE PHASE RESINS

Normal phase chromatography is used to separate fairly nonpolar compounds. Generally, the order of elution is that nonpolar compounds elute first and others follow based on increasing order of polarity. Conversely, reverse phase chromatography can be used to separate compounds that are much more polar. The more polar compounds elute first. The resins used for reverse phase are often macroporous beads. These are in the form of polymeric material, with high surface area and porosity, offering increased solute retention, selectivity, and superior loading capacity. An example is a highly cross-linked polystyrene resin ~50 Å pore diameter, such as XAD resins (see "Glossary").

2.7 HIGH-PERFORMANCE LIQUID CHROMATOGRAPHY

For high-performance liquid chromatography (HPLC) using modern macroporous resins as the stationary phase, the beads can be as small as 2.5 to 3.0 microns. Popular resin-based packing material is made from the copolymerization of polystyrene and divinylbenzene. The degree of cross-linking determines the rigidity of the resin. The advantages of this system include fast separations using high pressure and small volumes of solvent. Good resolution can be achieved with baseline separation using ultra high-performance technology.

C18 reverse phase packing materials have served as the *workhorse* resins of liquid chromatography for many years. There are now many different packing materials to choose from, ranging from C2 to C30. One manufacturer provides a C30 alkyl silane coated on spherical, porous silica gel. This LC column is designed to provide high shape selectivity for separation of hydrophobic structurally related isomers and compatible with highly aqueous mobile phases. Typically, reverse phase columns are available in 3 and 5 μm particle sizes and 4.6, 3.0, and 2.1 mm diameters, with an average pore diameter of 200 Å, ideal for separating hydrophobic structurally related isomers. Applications include separations of the long chain, structurally related isomers (e.g., carotenoids, steroids, etc.).

2.8 CAPILLARY ELECTROPHORESIS

Capillary electrophoresis (CE), also called capillary zone electrophoresis (CZE), is used to separate ionic species by charge, frictional forces, and hydrodynamic radius. In traditional electrophoresis, electrically charged analytes move in a conductive liquid medium under the influence of an electric field. CE separates species based on the size:charge ratio in the interior of a small capillary, filled with an electrolyte and shown in Figure 2.4.

2.9 POLYAMIDE GEL CHROMATOGRAPHY

The porous polyamide resin possesses hydrogen bond acceptor properties and is a very good medium for the separation of polyphenolic solutes such as phenolic acids, flavanols, and flavonoids. Typical solvent mixtures include aqueous acetic acid and ethanolic solutions.

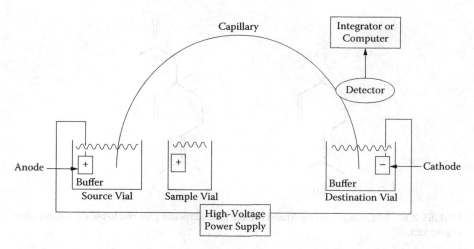

FIGURE 2.4 Capillary zone electrophoresis (CZE) is used to separate ionic species on the basis of charge, frictional forces, and hydrodynamic radii.

2.10 SIZE-EXCLUSION CHROMATOGRAPHY

In size-exclusion chromatography (SEC), molecules in solution are separated according to their size (i.e., molecular weight). This chromatography has been well adapted to the separation of large molecules, particularly proteins and industrial polymers. The chromatography is considered a form of gel filtration when an aqueous solution is used to transport the sample through the column. Gel-permeation chromatography occurs when an organic solvent is chosen as the mobile phase. A popular gel known as Sephadex LH20 is widely used to separate small and medium size phenolic compounds. An example of the separation of typical phenolic compounds on Sephadex LH20 is shown in Figure 2.5. Conversely, the LH-20 gel can be used as a filtration method whereby unwanted phenolic compounds can be retained on the gel, allowing the desired compounds of interest to pass through.

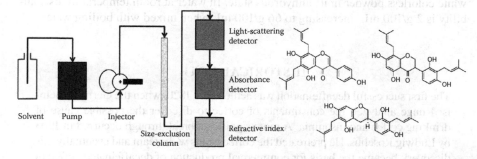

FIGURE 2.5 Typical setup used for size-exclusion (Sephadex LH-20) chromatography for the separation of phenolic compounds.

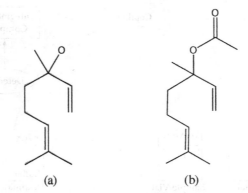

(a)   (b)

FIGURE 2.6   (a) Linalool and (b) linalyl acetate monoterpenes give rise to the distinctive odor of lavender.

EXAMPLE 2.1: ESSENTIAL OILS—HYDROSOL FROM LAVENDER

Lavandula angustifolia is the most widely cultivated species. It is a plant native to the western Mediterranean, but widely cultivated all over the world, including the British Isles, France, United States, Argentina, and Japan. Distillation of the lavender buds, spikes, and flower tips can produce the essential oil. During steam distillation of essential oils, large quantities of aqueous distillate are produced. The resulting mixture of essential oils, known as hydrosol, is widely used in aromatherapy and massage therapy. The distinctive odors of lavender are due to the key monoterpenes: linalool and linalyl acetate (Figure 2.6).

EXAMPLE 2.2: CAFFEINE AND THE DECAFFEINATION PROCESS OF COFFEE

Coffee comes from the *Coffea* genus of flowering plants. Although it is not responsible for the well-known coffee aroma, caffeine, present in the seeds, can protect the seeds themselves due to its toxicity to herbivores. One of the most popular species of the genus whose seeds contains caffeine is *Coffea arabica* (Figure 2.7).

The chemical formula of caffeine is $C_8H_{10}N_4O_2$ (Figure 2.8). It is a weakly basic, white colorless powder in its anhydrous state. In water at room temperature its solubility is 2 g/100 mL, increasing to 66 g/100 mL when mixed with boiling water.

HISTORICAL NOTE

The first successful decaffeination was achieved in 1820, when the German chemist Runge analyzed the constituents of coffee to discover the possible cause of drinking coffee and insomnia. A more significant breakthrough occurred in 1903 by Ludwig Roselius. He pretreated the coffee beans with steam and eventually this discovery became the basis for commercial production of decaffeinated coffee in the early 20th century.

FIGURE 2.7 Coffee berries from *Coffea arabica*. (Courtesy of Steven Foster.)

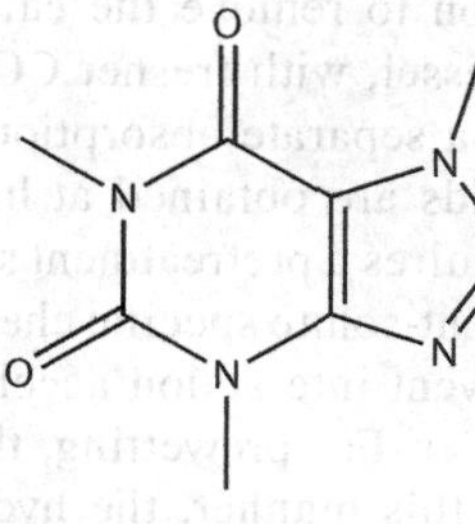

FIGURE 2.8 The chemical structure of caffeine.

There are three main ways to remove caffeine from coffee:

Extraction Procedure I: Solvent extraction using dichloromethane (DCM).
First, ground coffee in aqueous sodium carbonate is refluxed for 20 min-
utes; the mixture is filtered and cooled. The aqueous filtrate is partitioned
into DCM and this process can be repeated several times to extract more
caffeine. The addition of sodium carbonate converts the protonated form of
the caffeine, which is naturally present in coffee, to its free caffeine form
and this reaction is outlined in Figure 2.9.

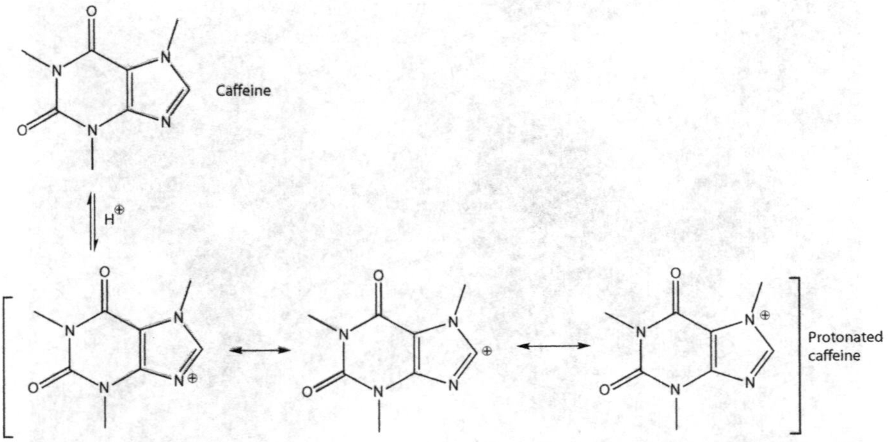

FIGURE 2.9 Caffeine extraction: The addition of sodium carbonate converts the protonated form of the caffeine back to its free form.

Extraction Procedure II: Supercritical carbon dioxide extraction. There are advantages to this method: eliminating the use of flammable and toxic solvents, and the caffeine is easily removed from the final product. In this process, the caffeine diffuses into supercritical CO_2 with water. In a continuous extraction to remove the caffeine, the beans enter at the top of an extractor vessel, with fresher CO_2 entering at the bottom. Recovery is achieved in a separate absorption chamber with water. In this manner, greater yields are obtained at higher pressures and temperature. The process requires a pretreatment step. The addition of polar cosolvents affects cosolvent-solute specific chemical or physical interactions. The solvent–cosolvent interaction accelerates the extraction and makes the extraction easier. For prewetting, the material is humidified with ultrapure water. In this manner, the hydrogen bonds linking the caffeine to its natural matrix are ruptured. Cell membrane swelling enhances solute diffusion. Subsequently, the quality of caffeine can reach a purity >94%, and this is usually the acceptable criteria for use in the soft drink and drug industries.

Extraction Procedure III: Activated charcoal. There are some advantages to using charcoal: it is economical, "green," and easily regenerated by heat and steam. By choosing activated carbon with the appropriate numbers of micropores and a surface area up to 1000 m^2 per gram, it is possible to achieve good performance of adsorption. Cleaned green coffee beans are first soaked in water, and the caffeine and other soluble content will migrate to the aqueous phase. The water is filtered through the active charcoal and only caffeine will continue to migrate in water. These recovered and dried coffee beans are now decaffeinated.

SUMMARY

This chapter provides examples of separations of natural products from various sources, various chromatographic techniques, and extraction methods. Further downstream, isolation techniques applicable to large-scale and commercial production are shown.

QUESTIONS

1. Previously, organic solvents such as benzene, chloroform, carbon tetrachloride, and hexane were very popular choices. Why are they no longer used?
2. What are the most common solvents or combinations that are used in the natural products industry today?
3. Using the answers for Question 2, why are they chosen?
4. Name any two of the four first extraction techniques.
5. Name the ways to separate polar compounds and give some examples.

FURTHER READING

1. R. Cooper, A. C. Horan, F. Gentile, V. Gullo, D. Loebenberg, J. Marquez, M. Patel, M. S. Puar, and I. Truumees. 1988. Sch 37137, a novel antifungal compound produced by a *Micromonospora* sp. taxonomy, fermentation, isolation, structure and biological properties. *J Antibiot (Tokyo)* 41(1):13–19.

GAS CHROMATOGRAPHY

P. J. Marriot, R. Shellie, and C. Cornwell. 2001. Gas chromatographic technologies for the analysis of essential oils. *J Chromatogr A* 936(1–2):1–22.

POLYAMIDE

M. Gao, X. Wang, M. Gu, Z. Su, Y. Wang, and J.-C. Janson. 2011. Separation of polyphenols using porous polyamide resin and assessment of mechanism of retention. *Journal of Separation Science* 34(15):1853–1858.

ESSENTIAL OILS

K. L. Adam. 2006. Lavender Production, Products, Markets, and Entertainment Farms. ATTRA–National Sustainable Agriculture Information Service.

E. J. Bowles. 2004. *The Chemistry of Aromatherapeutic Oils*. 3rd ed. Allen & Unwin.

R. Tisserand and T. Balacs. 1995. *Essential Oil Safety: A Guide for Health Care Professionals*. 1st ed. Churchill Livingstone.

CAFFEINE AND SUPERCRITICAL CO$_2$ EXTRACTION

J. Tello, M. Viguera, and L. Calvo. 2011. Extraction of caffeine from Robusta coffee (*Coffea canephora* var. Robusta) husks using supercritical carbon dioxide. *Journal of Supercritical Fluids* 59:53–60.

CHARCOAL

K. S. W. Sing. 2005. Overview of physical adsorption by carbons. In *Adsorption by Carbons*, edited by E. Bottani and J. Tascón. 1st ed. Elsevier Science.

3 Structure Elucidation

Elucidation of chemical structures requires piecing together a combination of chemical and spectroscopic information. Typically, the spectroscopic tools most frequently used are nuclear magnetic resonance (NMR), which includes ^{1}H- and ^{13}C-NMR; ultraviolet (UV); infrared (IR); and mass spectrometry (MS). Use of circular dichroism (CD) and X-ray crystallography are often used in support of determining the stereochemistry and the absolute configuration. Early studies on the structural elucidation of a natural product required large amounts of material, preferably recrystallized to homogeneity. The need for a sharp melting point and elemental analysis were paramount. This information was followed by a combination of degradation studies and derivatization reactions, and eventually chemical synthesis to confirm the structure. With the advent of modern-day spectroscopic techniques, microgram amounts are often sufficient to complete the determination of structure using only spectroscopic techniques.

3.1 NUCLEAR MAGNETIC RESONANCE

Only nuclei with spin number $I \neq 0$ can absorb/emit electromagnetic radiation. If a nucleus has an even mass A and even charge Z, then the nuclear spin I is zero. Examples are ^{12}C, ^{16}O, and ^{32}S. However, a nucleus with a mass $I = n/2$, where n is an odd integer, for example, ^{1}H, ^{13}C, ^{15}N, and ^{31}P, in which case it is possible to detect their nuclear magnetic resonance (NMR) signals. The chemical shift of the nucleus is the difference between its resonance frequency and a standard. Usually, this quantity is reported in parts per million (ppm) and given the symbol delta, δ. In NMR spectroscopy as illustrated in Figure 3.1 for the simple compound ethanol, this standard is often tetramethylsilane, $Si(CH_3)_4$, abbreviated as TMS. In the ethanol

HISTORICAL NOTE

In the 20th century, as improvements continued in separation techniques and more compounds were isolated from plant and microbial sources, there were quantum leaps in the spectroscopic techniques used to determine chemical structures. An anecdotal story made the rounds in the United Kingdom in the early 1960s, when a Nobel Laureate, sitting on the panel of a candidate's Ph.D. thesis defense, challenged the student to show he had determined that his isolated chemical compound was pure. All the spectral techniques were described in detail to support the purity and to determine the chemical structure. Alas, the candidate was told to come back and redefend his thesis when he had taken the melting point.

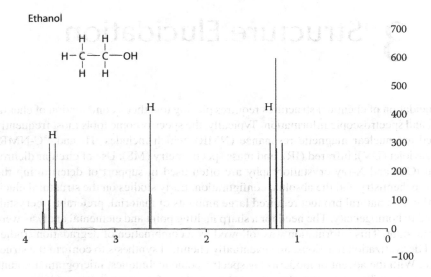

FIGURE 3.1 ^1H NMR spectrum of ethanol.

molecule there are three distinct types of hydrogen: three on one carbon, two on another, and the proton of the OH group. The chemical shifts are different and possess unique positions in the NMR spectrum. The proton signals are observed as multiple peaks due to the "splitting" by the corresponding adjoining protons. In this example, the CH_3 protons are observed as a triplet and the CH_2 group as a quartet. The OH proton is not affected by any adjoining proton and remains as a singlet.

3.2   ULTRAVIOLET AND INFRARED SPECTROSCOPY

Ultraviolet (UV) light is electromagnetic radiation with a wavelength shorter than that of visible light but longer than X-rays, in the range 10 to 400 nm, but the frequencies are invisible to humans. In chemical analysis and for separations, the use of a UV or photodiode array (PDA) detector based on detection of the molecule's UV spectrum is a useful tool. In particular, phenolic compounds and conjugated systems possess strong, easily characterized chromophores. Examples of UV spectra of distinctive chromophores of select carotenoids are shown in Chapter 11.

Infrared (IR) spectroscopy (Figure 3.2) measures the infrared region of the electromagnetic spectrum light with a longer wavelength and lower frequency than visible light.

Examples of intense IR stretches in the region of 1660–1780 cm^{-1} characterize various C=O stretching bands. Examples of molecules possessing these characteristics are the β-lactam antibiotics described in Chapter 9.

3.3   MASS SPECTROMETRY

Mass spectrometers are used to measure the difference in mass-to-charge ratio (m/z or m/e) of ionized atoms, molecules, and fragment ions. These charged ions

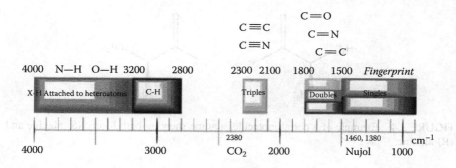

FIGURE 3.2 IR stretching frequency ranges.

are separated from each other and the mass spectrum records the quantity of ions of particular mass-to-charge ratios. Peak heights recorded on a mass spectrum are proportional to the number of ions of each mass. In this manner, the readout using mass spectroscopy is used to determine the molecular weight and help support the identity and features of the molecular structure.

As an example, the mass spectrum of a simple alkane, pentane, C_5H_{12} is presented in Figure 3.3. The highest mass peak, m/z 72 mu corresponds to the molecular mass of the pentane molecule. Subsequent fragmentation yields smaller fragment ions, which are identified in Table 3.1.

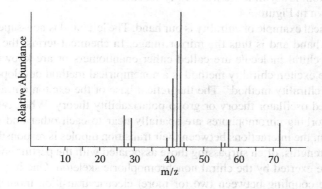

FIGURE 3.3 Mass spectrum of pentane, showing fragment ions.

TABLE 3.1
Mass Spectrum of Pentane

m/z	Fragment	Loss of
72	CH₃CH₂CH₂CH₂CH₃	
57	·CH₂CH₂CH₂CH₃	CH₃ (m-15)
43	·CH₂CH₂CH₃	CH₂ (m-14)
29	·CH₂CH₃	CH₂ (m-14)

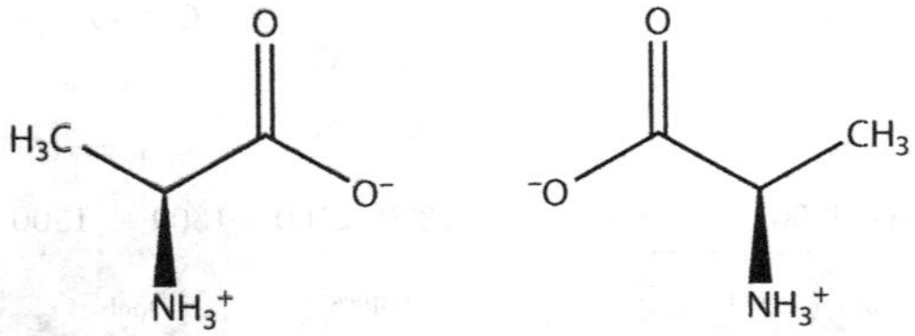

FIGURE 3.4 Example of a chiral molecule, alanine showing both forms: (S)-alanine and (R)-alanine.

3.4 CIRCULAR DICHROISM

Circular dichroism (CD) refers to the differential absorption of left and right circularly polarized light. It is exhibited in the absorption bands of optically active chiral molecules. CD spectroscopy has a wide range of applications in many different fields. Most notably, UV CD is used to investigate the secondary structure of proteins and can allow a measure of the differences in the absorption of left-handed polarized light versus right-handed polarized light, which arise due to structural asymmetry. The absence of a regular structure results in zero CD intensity, whereas an ordered structure results in a spectrum, which can contain both positive and negative signals.

A chiral molecule has a non-superimposable mirror image, due to the fact that the carbon atom is asymmetric when fully substituted. An example is the amino acid alanine shown in Figure 3.4.

The classical example of chirality is our hand. The left hand is non-superimposable on the right hand and is thus the mirror image. In chemical terms, the two mirror images of a chiral molecule are called either enantiomers or are known as optical isomers. The exciton chirality method is a nonempirical method developed from the "dibenzoate chirality method." The theoretical base of the exciton chirality method is the coupled oscillator theory or group polarizability theory. When two (or more) strongly absorbing chromophores are spatially near to each other and constitute a chiral system, the interactions between their transition dipoles is responsible for large rotational strengths, often surpassing those associated with the perturbations on each chromophore exerted by the chiral nonchromophoric skeleton. One important interaction is the coupling between two (or more) electric transition moments (exciton coupling), which results in the splitting of the excited energy level. The sign of exciton chirality determines the stereochemistry. An example is shown in Figure 3.5.

3.5 X-RAY CRYSTALLOGRAPHY

X-ray crystallography, illustrated in Figures 3.6 and 3.7, is used for determining the arrangement of atoms within a crystal, in which a beam of X-rays strikes the crystal and causes the beam of light to spread into many specific directions. From the angles and intensities of these diffracted beams, a crystallographer can produce a three-dimensional picture of the density of electrons within the crystal. From this electron

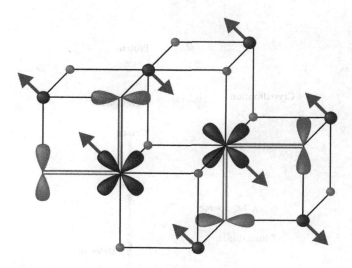

FIGURE 3.5   Example of a circular dichroism (CD) spectrum.

density, the mean positions of the atoms in the crystal can be determined, as well as their chemical bonds, their disorder, and various other information.

In an X-ray diffraction measurement, a crystal is mounted on a goniometer and gradually rotated while being bombarded with X-rays, producing a diffraction pattern of regularly spaced spots known as *reflections*. The two-dimensional images taken at different rotations are converted into a three-dimensional model of the density of electrons within the crystal using the mathematical method of Fourier transforms, combined with chemical data known for the sample. Poor resolution (fuzziness) or even errors may result if the crystals are too small or not uniform enough in their internal makeup. Producing good quality crystals is still recognized as an art!

HISTORICAL NOTE

Wilhelm Conrad Röntgen discovered X-rays in 1895, just as the studies of crystal symmetry were being concluded. The use of X-rays created the X-ray diffraction pattern of crystals. The study of X-ray crystallography of biological molecules accelerated with the contributions from Dorothy Crowfoot Hodgkin, who solved the structures of cholesterol (1937), vitamin B_{12} (1945), and penicillin (1954), for which she was awarded the Nobel Prize in Chemistry in 1964. In 1969, she succeeded in solving the structure of insulin on which she worked for over 30 years.

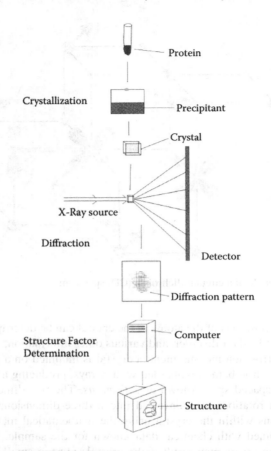

FIGURE 3.6 Overview of X-ray crystallography.

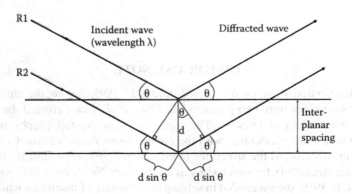

FIGURE 3.7 A set of parallel planes with interplanar spacing d produces a diffraction beam when X-rays R1 and R2 are reflected by the planes at an angle θ. If the difference in path length, 2(d sin θ), is equal to an integral number (n) of wavelengths, then they constructively interfere with each other. This is Bragg's law of diffraction.

EXAMPLE 3.1: PARNAFUNGINS

An example of the use of combined spectroscopic techniques is nicely illustrated with the discovery of the parnafungins. These compounds were discovered as new microbial metabolites from a fungal species. The parnafungins are actually a mixture of four closely related compounds—four distinct stereoisomers named A1, A2, B1, and B2 (see Figure 3.8).

The structure determination presents some challenges. The compounds are unstable and form equilibrium mixtures. The parnafungins contain a xanthone ring system (UV and NMR are more complex) and possess a rare structural feature in nature: cyclic: N-O-CO-, an isoxazolidinone ring (Figure 3.9).

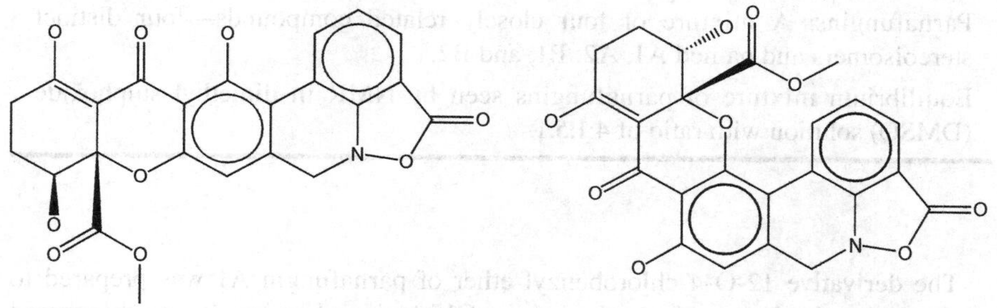

FIGURE 3.8 Chemical structures of the parnafungin group of compounds. (From Parish, S. K. et al., 2008, *J Am Chem Soc* 130(22):7060–7066. With permission from the American Chemical Society.)

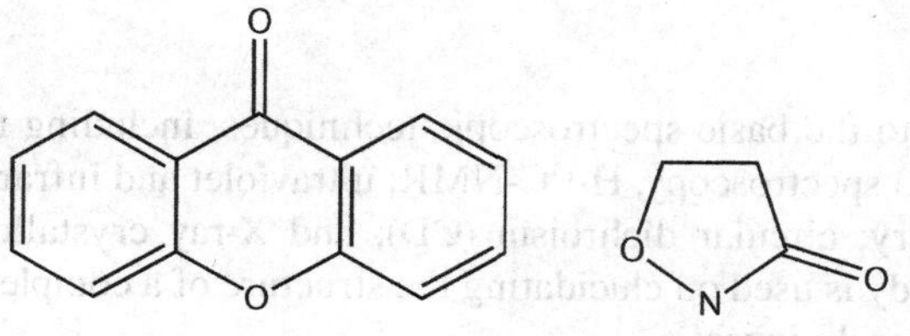

FIGURE 3.9 Isoxazolidinone ring: Key fragments of the parnafungin molecule used to confirm the structure. (From Parish, S. K. et al., 2008, *J Am Chem Soc* 130(22):7060–7066. With permission from the American Chemical Society.)

Case Study

Fermentation of a strain of *Fusarium larvarum*

Isolation of whole broth into acetone, breaks the cells then filtration, retain acetone, concentrate to remove acetone, load aqueous liquid on resin

Reverse phase, CHP20P chromatography, elute aq. acidic-acetonitrile

Elution-extrusion countercurrent chromatography (EECCC), elute mobile phase 1:1:1:1 hexane/ethyl acetate/methanol/water

Parnafungins: A mixture of four closely related compounds—four distinct stereoisomers and named A1, A2, B1, and B2.

Equilibrium mixture of parnafungins seen by NMR in dimethyl sulphoxide (DMSO) solution with ratio of 4:1:5:1.

The derivative 12-O-4-chlorobenzyl ether of parnafungin A1 was prepared to establish the absolute configuration of the C15-hydroxyl group. It was discovered that parnafungin A1 shows the syn configuration between the 15-hydroxyl and the neighboring 15a-methyl carboxylate. The complete X-ray structure of the derivative is displayed in Figure 3.10. Since this was an unusual compound, X-ray analysis offered (1) absolute proof of structure and (2) resolved the stereo centers—from relative to absolute stereochemistry. It should be noted that to gain useful information using X-ray crystallography, it is necessary to obtain good quality crystals from a pure stable derivative. Usually, the derivative contains a large atom (e.g., chlorine or bromine atom).

SUMMARY

An introduction to the basic spectroscopic techniques, including nuclear magnetic resonance (NMR) spectroscopy, H-^{13}C-NMR; ultraviolet and infrared spectroscopy; mass spectrometry; circular dichroism (CD); and X-ray crystallography are provided. A case study is used on elucidating the structure of a complex molecule using a combination of techniques.

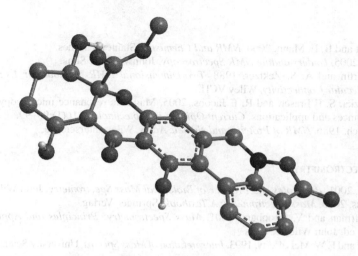

FIGURE 3.10   X-ray crystallographic structure of the parnafungin derivative. (From Parish, S. K. et al., 2008, *J Am Chem Soc* 130(22):7060–7066. With permission from the American Chemical Society.)

QUESTIONS

1. What are the main spectroscopic techniques used for structure determination?
2. Which nuclei lend themselves to NMR?
3. Why is the methylene in ethanol (CH₃CH₂OH) observed as a quartet in the H-NMR spectrum?
4. Why is the quartet seen as a ratio of 1:3:3:1?
5. What do you understand about the term *500MHz H-NMR spectrum*?
6. What other sized instruments are available? What are the benefits of each one?
7. Why is TMS used as an internal standard for running NMR spectra?
8. In mass spectrometry, what do you understand about the terms *positive ion mode* and *negative ion mode*?
9. In X-ray crystallography the compound of interest is often derivatized with the addition of a group containing a "heavy" atom. Give an example and explain the reason.

FURTHER READING

1. C. A. Parish, S. K. Smith, K. Calati, D. Zink, K. Wilson, T. Roemer, B. Jiang, et al. 2008. Isolation and structure elucidation of parnafungins, antifungal natural products that inhibit mRNA polyadenylation. *J Am Chem Soc* 130(22):7060–7066.

3. Why is the methylene in ethanol (CH_3CH_2OH) observed as a quartet in the H-NMR spectrum?

NMR

J. W. Akitt and B. E. Mann. 2000. *NMR and Chemistry.* Stanley Thornes.

J. Keeler. 2005. *Understanding NMR Spectroscopy.* John Wiley & Sons.

G. E. Martin and A. S. Zektzer. 1988. *Two-Dimensional NMR Methods for Establishing Molecular Connectivity.* Wiley VCH.

J. M. Tyszka, S. E Fraser, and R. E Jacobs. 2005. Magnetic resonance microscopy: Recent advances and applications. *Current Opinion in Biotechnology* 16(1):93–99.

K. Wuthrich. 1986. *NMR of Proteins and Nucleic Acids.* Wiley Interscience.

MASS SPECTROMETRY

D. Chabil. 2001. *Principles and Practice of Biological Mass Spectrometry.* John Wiley.

J. H. Gross. 2006. *Mass Spectrometry: A Textbook.* Springer-Verlag.

E. de Hoffman and V. Stroobant. 2001. *Mass Spectrometry: Principles and Applications.* 2nd ed. John Wiley & Sons.

F. Tureček and F. W. McLafferty. 1993. *Interpretation of Mass Spectra.* University Science Books.

X-RAY CRYSTALLOGRAPHY

L. Bragg. 1968. X-ray crystallography. *Sci Am* 219:58–70.

R. Diamond. 1985. Diffraction methods for biological macromolecules. Real space refinement. *Methods Enzymol* 115:237–252.

A. McPherson. 2003. *Introduction to Macromolecular Crystallography.* John Wiley & Sons.

G. Rhodes. 2000. *Crystallography Made Crystal Clear.* Academic Press.

CIRCULAR DICHROISM

P. Atkins and J. de Paula. 2005. *The Elements of Physical Chemistry.* 4th ed. W. H. Freeman.

N. Berova and K. Nakanishi. 2000. Exciton chirality method: Principles and applications. In *Circular Dichroism: Principles and Applications*, edited by N. Berova, K. Nakanishi, and R. W. Woody. 2nd ed. Wiley-VCH.

G. D. Fasman. 1996. *Circular Dichroism and the Conformational Analysis of Biomolecules (Siberian School of Algebra and Logic).* Springer.

K. Nakanishi and N. Berova. 1994. *Circular Dichroism: Principles and Applications.* Vch Pub.

E. I. Solomon and A. B. P. Lever. 2006. *Inorganic Electronic Structure and Spectroscopy: Applications and Case Studies.* Vol. II. Wiley-Interscience.

4 Isomers and Building Blocks

Stereochemistry is a term used to describe the *relative* (and absolute) orientation of the groups within a molecule. The stereochemistry of carbon–carbon bonds represented by *cis/trans* isomerism is also known as geometric isomerism. The terms *cis* means "on the same side" and *trans* means "across." Double bonds in unsaturated compounds are represented either in the *cis* or *trans* configuration (Figure 4.1). An example of a naturally occurring terpene, occurring in two isomeric forms, highlights the stereochemical differences of *cis* and *trans* isomers, shown in Figure 4.2. Although these isomers of the compound, named β-ocimene, have identical molecular formulas, they can be chromatographically separated and they have different spectral properties.

An alternative use of the *cis* and *trans* notation to indicate relative stereochemistry is the *E-Z* notation. A set of rules known as the Cahn-Ingold-Prelog rules gives each substituent on a double bond an assigned priority. In the example shown in Figure 4.3, as an extension of the use of the *cis* and *trans* notation to indicate relative stereochemistry, the *E-Z* notation is generally accepted to describe the absolute configuration of the double bonds. The higher priority is given to the methyl group. Thus, when the two groups of higher priority are on opposite sides of the double bond, the bond is assigned the configuration *E*. Conversely, if the two groups are on the same side of the double bond, the bond is assigned the configuration *Z* (Figure 4.3).

4.1 TERPENES AND THE ISOPRENE RULE

Terpenes constitute the largest and most diverse class of natural products and are found mostly in plants. Terpenes are made up of isoprene (isopentane) units: a five carbon subunit from which the isoprene rule is derived. Terpenes are thus identified by tracing the 5n carbon atoms in the molecule. However, it should be noted that larger and more complex terpenes (e.g., squalene [Figure 4.4] and lanosterol) are also found in animals.

Monoterpenoids represent a large class of terpene compounds. Biochemical modifications such as oxidation or a rearrangement reaction produce a variety of open chain and cyclized monoterpenoids; and several structures are shown in Figure 4.5. Monoterpenes consist of two isoprene units and have the general molecular formula $C_{10}H_{16}$. They are low molecular weight compounds, are very volatile, and many can be recognized by their distinctive odors.

Isoprenes are made from C-5 units possessing 5n carbon atoms; n is an integer. Two or more isoprene molecules are linked to one another to create terpenoids. Linking between two isoprene molecules can occur in three ways, either with the "head" or "tail" of the molecule, as shown in Figure 4.6.

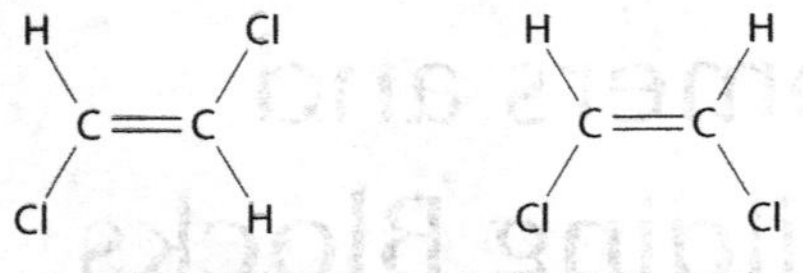

trans-1,2-dichloroethene　　　cis-1,2-dichloroethene

FIGURE 4.1　Example of *cis* (same) versus *trans* (opposite) isomers in a molecule.

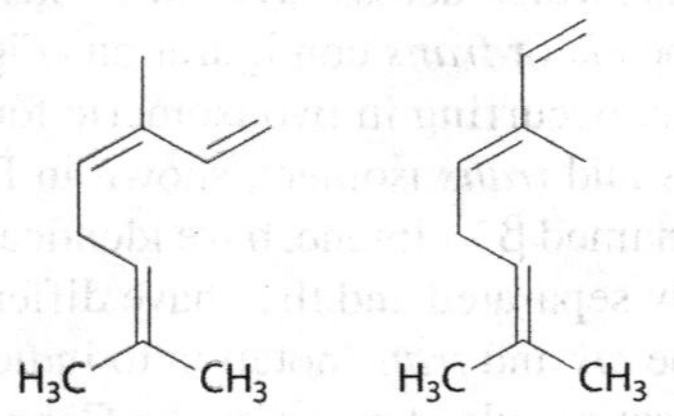

cis　　　　　　　trans

FIGURE 4.2　Structures of the two isomers of the terpene β-ocimene in *cis* and *trans* forms.

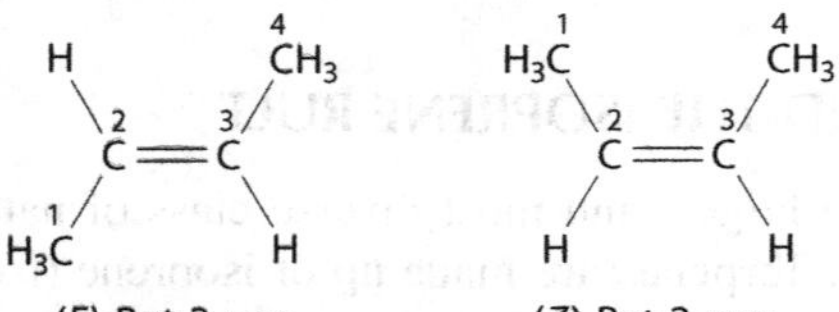

(*E*)-But-2-ene　　　　　(*Z*)-But-2-ene

FIGURE 4.3　E, Z isomers of but-2-ene.

FIGURE 4.4　Chemical structure of the naturally occurring terpenoid squalene.

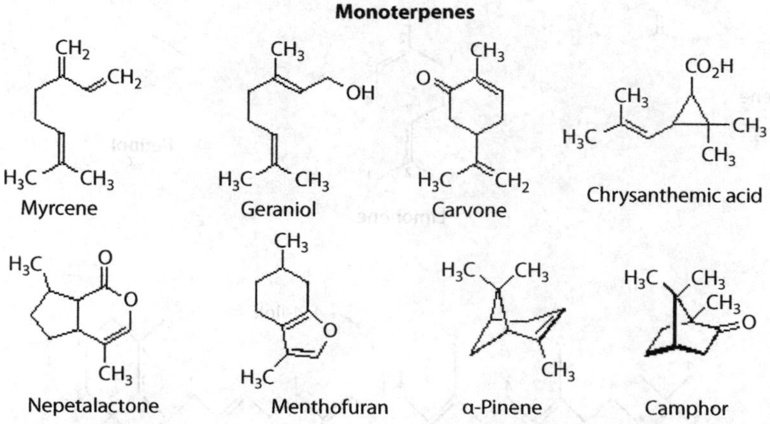
Monoterpenes
CH2
CH2
H3C
CH3
Myrcene
CH3
OH
H3C
CH3
Geraniol
CH3
O
H3C
CH2
Carvone
CO2H
CH3
H3C
CH3
CH3
Chrysanthemic acid
H3C
O
O
CH3
Nepetalactone
CH3
O
H3C
Menthofuran
H3C
CH3
CH3
α-Pinene
H3C
CH3
CH3
O
Camphor

FIGURE 4.5 Examples of naturally occurring oxygenated monoterpene structures.

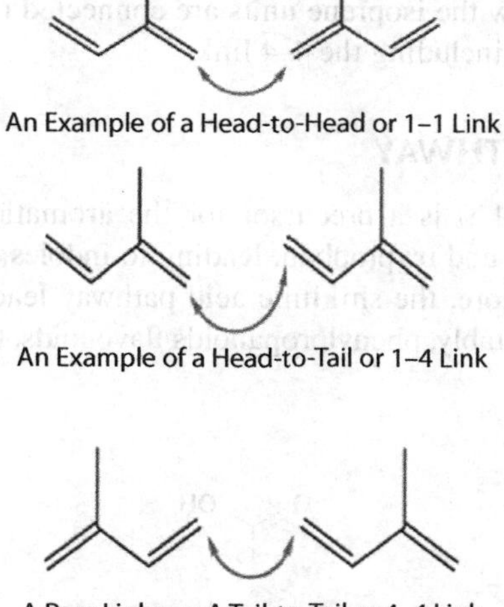
An Example of a Head-to-Head or 1–1 Link
An Example of a Head-to-Tail or 1–4 Link
A Rare Linkage: A Tail-to-Tail or 4–4 Link

FIGURE 4.6 Different ways of connecting isoprene units.

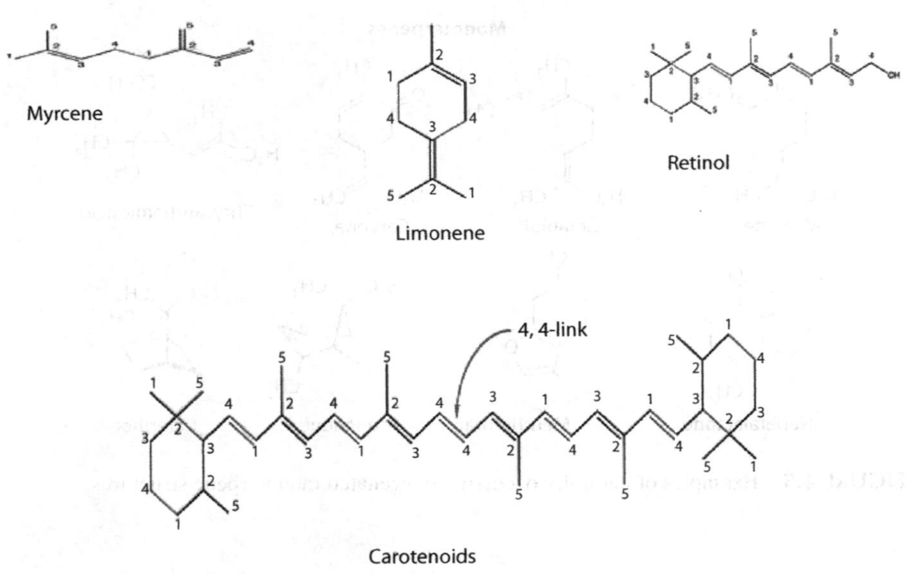

Myrcene

Limonene

Retinol

4, 4-link

Carotenoids

FIGURE 4.7 Examples of various terpenoid molecules found in nature.

Some more naturally occurring terpenes are varied in structure; the complexity and level of importance are indicated in Figure 4.7. They are formed by the linking of isoprene units and eventually can combine to form naturally occurring molecules known as the carotenoids, which will be discussed further in Chapter 11. This type of terpene illustrates how the isoprene units are connected to form long chain poly-unsaturated compounds including the 4-4 link.

4.2 SHIKIMATE PATHWAY

Shikimic acid (Figure 4.8) is a precursor for the aromatic amino acids, such as phenylalanine, tyrosine, and tryptophan, leading to indoles, indole-like derivatives, and alkaloids. Furthermore, the shikimic acid pathway leads to a large number of phenolic compounds, notably, phenylpropanoids flavonoids, tannins, and lignins discussed in Chapter 6.

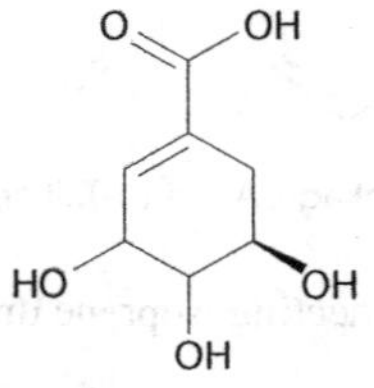

FIGURE 4.8 Structure of shikimic acid.

The metabolic route used by bacteria, fungi, algae, parasites, and plants for the biosynthesis of aromatic amino acids is the shikimate pathway. This pathway is not found in animals or in humans, hence the products of this pathway represent essential amino acids that must be obtained from the diet. The first enzyme involved is shikimate kinase, an enzyme that catalyzes the ATP-dependent phosphorylation of shikimate to form shikimate 3-phosphate. Shikimate 3-phosphate is then coupled with phosphoenolpyruvate to give 5-enolpyruvylshikimate-3-phosphate via the enzyme 5-enolpyruvylshikimate-3-phosphate (EPSP) synthase as shown in Figure 4.9, where the entire shikimate pathway is outlined. The 5-enolpyruvylshikimate-3-phosphate is then transformed into chorismate by a chorismate synthase. The resulting prephenic acid is synthesized by a rearrangement of chorismate. The prephenate is oxidatively decarboxylated with retention of the hydroxyl group to give *p*-hydroxyphenyl-pyruvate, which in turn undergoes transamination using glutamate as the nitrogen source, to give two compounds: tyrosine and alpha keto-glutarate.

4.3 MEVALONATE PATHWAY

The mevalonate pathway is another important cellular pathway present in all higher eukaryotes and many bacteria. It is important for the production of dimethyl-allyl pyrophosphate (DMAPP) and isopentenyl pyrophosphate (IPP). These two compounds serve as the basis for the biosynthesis of molecules used in processes as diverse as terpenoid synthesis, protein prenylation, cell membrane mainte-nance, hormones, protein anchoring, and N-glycosylation. They are also a part of steroid biosynthesis.

4.4 POLYKETIDES

Polyketides are secondary metabolites generally derived from bacteria, fungi, plants, and animals. These compounds form a large class of diverse compounds, which are characterized by more than two carbonyl groups connected by single intervening carbon atoms. Polyketides are usually biosynthesized through the decarboxylative condensation of malonyl-CoA derived and extended units. The chemical structure of the malonate ion is shown in Figure 4.10, created by a process known as Claisen condensation. The condensation reaction allows for the formation of carbon–carbon bonds. The reaction occurs between two esters (or in some cases between a carbonyl and ester). When the reaction takes place under basic conditions a β-keto ester (or a β-diketone) is formed.

Structurally, this diverse family of natural products possesses various biological activities and pharmacological properties. There are numerous examples of polyketide antibiotics, antifungals, and other biologically actives in commercial use (e.g., erythromycin and tetracyclines are represented in Chapter 9) and they are broadly divided into three classes:

Type I polyketides (e.g., macrolides)
Type II polyketides (e.g., aromatic molecules)
Type III polyketides (often small aromatic molecules produced by fungal species)

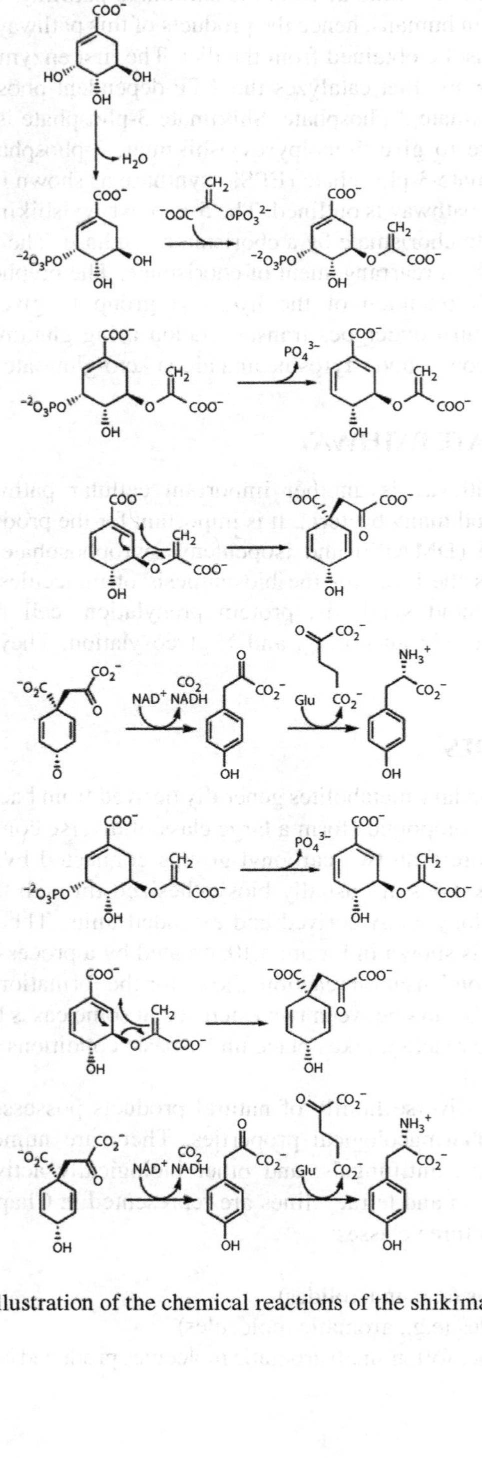

FIGURE 4.9 The illustration of the chemical reactions of the shikimate pathway.

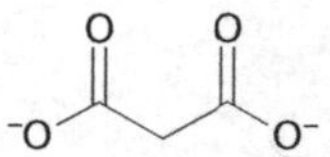

FIGURE 4.10 Chemical structure of the malonate ion.

SUMMARY

The concept of stereochemistry and isomers in organic chemistry is presented leading to an introduction to isomers of double bonds. The chapter describes the core building blocks and pathways for natural products including the isoprene rule, shikimic acid as a precursor to aromatic compounds, mevalonate as a precursor to terpenoids, and polyketides as a precursor to alkaloids.

QUESTIONS

1. There are two types of metabolites. What are they?
2. Give an example of each type of metabolite.
3. Give two examples each of molecules derived from (a) isoprenes, (b) shikimate, (c) mevalonate, (d) polyketide, and (e) malonate pathways.

FURTHER READING

P. M. Dewick. 2011. *Medicinal Natural Products: A Biosynthetic Approach.* Wiley.
S. P. Stanforth. 2006. *Natural Product Chemistry at a Glance.* Wiley-Blackwell.

Section II

Selected Classes
of Natural Products

5 Sugar and Fat and All of That

5.1 CARBOHYDRATES

Carbohydrates (carbon hydrates, $C_n(H_2O)_n$, hence their name), are the most abundant class of organic compounds found in living organisms. They originate as products of photosynthesis: condensation of carbon dioxide requiring light, energy, and the pigment chlorophyll.

$$n\ CO_2 + n\ H_2O + energy \rightarrow C_nH_{2n}O_n + n\ O_2$$

Carbohydrates, also known as polysaccharides or sugars, are a major source of metabolic energy, both for plants and for animals that depend on plants for food. They are a component of the energy transport compound ATP and are on recognition sites on cell surfaces. Importantly, sugars are also one of the three essential components of both DNA (e.g., ribose) and RNA (e.g., deoxyribose).

The simplest carbohydrate is a monomer (monosaccharide), represented either by a six- or five-membered ring structure, with the examples of glucose and ribose respectively, shown in Figure 5.1.

The six-membered ring sugar can be represented by several other forms designated by the structures shown in Figure 5.2. These forms of the monomer can be:

- As ketose sugars having a ketone function or an acetal equivalent
- As aldose sugars having an aldehyde function or an acetal equivalent

Examples of simple hexose and pyranose structures are shown in Figure 5.3. All compounds are isomers of glucose because the configurations of the OH group can vary on each of the carbons of the ring. The two rows represent common ways to draw the structures of the monomers.

5.2 VARIETY AND COMPLEXITY OF SUGARS

When these monosaccharides are linked they form polysaccharide chains through glycosidic bond linkages. Linking the sugars can be in a linear or branched manner, or as a combination of both ways, and various carbohydrates (e.g., disaccharides, oligosaccharides, and polysaccharides) can be generated. Polysaccharides can attain very high molecular weight ranges of over 100,000 daltons.

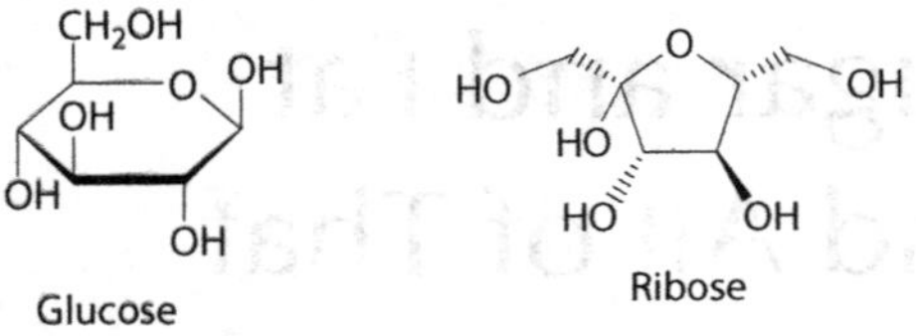

FIGURE 5.1   Chemical structures of two monosaccharides: glucose and ribose.

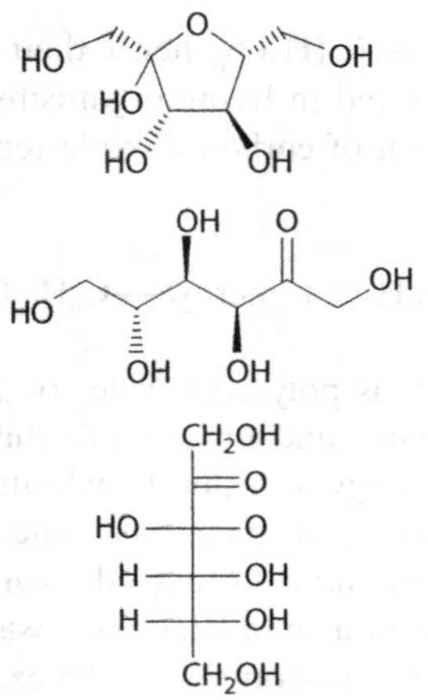

FIGURE 5.2   Alternative forms of the six-membered ring sugar.

5.3   FRUCTOSE AND GLUCOSE

Fructose is a simple sugar commonly found in fruits and vegetables. Glucose, also known as grape or blood sugar, is present in all major carbohydrates, for example, starch and table sugar. Although both are a good source of energy, excess of glucose can be fatal to diabetic patients, and excess of fructose in the human body can lead to health problems such as insulin resistance and liver disease.

5.4   THE DISACCHARIDE SUCROSE

There are two types of sugars: reducing and nonreducing. Monosaccharides are almost all reducing sugars, for example, glucose or lactose. Detection is achieved with the use of Fehling's reagent. In the Fehling reaction, reducing sugars should turn red and develop a red precipitate, which comes from the reduction of copper (II) ions to copper (I) oxide.

Conversely, a good example of a nonreducing sugar is sucrose (Figure 5.4). Sucrose is composed of glucose (left) and fructose (right), which are linked together

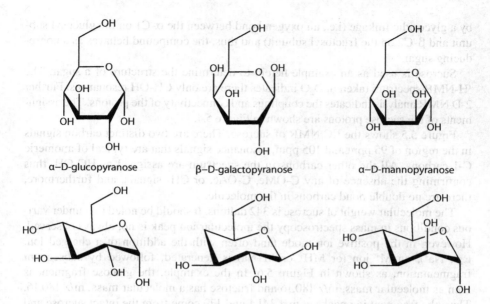

α–D-glucopyranose β–D-galactopyranose α–D-mannopyranose

FIGURE 5.3 Examples of some pyranose forms of hexose sugars.

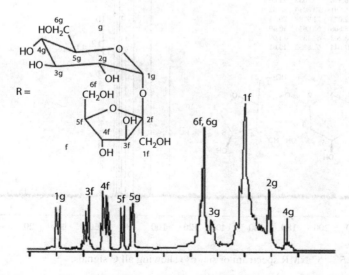

R =

FIGURE 5.4 Chemical structure and ¹H NMR spectrum of sucrose with chemical assignments.

by a glycosidic linkage (i.e., an oxygen bond between the α-C1 on the glucosyl subunit and β-C2 on the fructosyl subunit) and thus, the compound behaves as a nonreducing sugar.

Sucrose is used as an example herein to determine the structure of a sugar. The ¹H-NMR spectrum taken in D_2O indicates there are only CH-OH resonances. Further 2-D NMR analysis indicates the couplings and connectivity of the protons. The assignments of the sucrose protons are shown in Figure 5.4.

Figure 5.5 shows the ¹³C NMR of sucrose. There are two distinct carbon signals in the region of 95 ppm and 105 ppm, resonance signals that are typical of anomeric C-1 carbons. All the other carbons in the spectrum are assigned as HC-OH, thus confirming the absence of any C-OMe, C-OAc, or CH_3 signals, and furthermore, there are no double bond carbons in the molecule.

The molecular weight of sucrose is 342 daltons. It should be noted that under various conditions in mass spectroscopy the molecular ion peak is not always observed. However, in the positive ion mode (and often with the addition of a charged ion, e.g., Na^+), an M^+ ion (or MH^+ or M+Na) is generated, followed by subsequent fragmentation, as shown in Figure 5.6. In the example, the glucose fragment is seen as molecular mass, m/z 180.16 and fructose has a molecular mass, m/z 180.16. Thus, the fragment ion peaks at m/z 341.1 and 179 come from the intact sucrose and

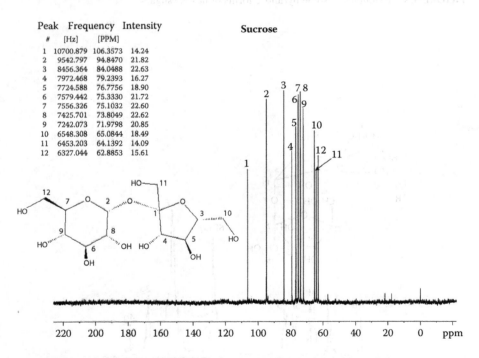

Peak	Frequency	Intensity
#	[Hz]	[PPM]	
1	10700.879	106.3573	14.24
2	9542.797	94.8470	21.82
3	8456.364	84.0488	22.63
4	7972.468	79.2393	16.27
5	7724.588	76.7756	18.90
6	7579.442	75.3330	21.72
7	7556.326	75.1032	22.60
8	7425.701	73.8049	22.62
9	7242.073	71.9798	20.85
10	6548.308	65.0844	18.49
11	6453.203	64.1392	14.09
12	6327.044	62.8853	15.61

Sucrose

FIGURE 5.5   ¹³C NMR spectrum of sucrose showing all C signals.

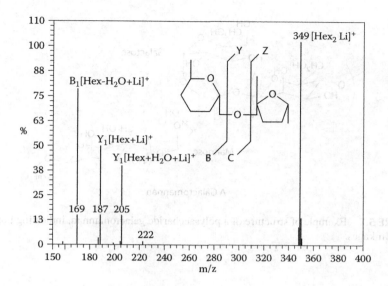

FIGURE 5.6 The mass spectrum of sucrose showing a lithiated pseudo-molecular ion and subsequent fragment ions.

glucose/fructose moieties, respectively. Other peaks represent small fragments of the monomer: ion peaks with mass m/z 89 and m/z 118 in the sucrose spectrum originate from further fragmentation of glucose; and m/z 101 and m/z 143 originate as fragment ions derived from fructose.

5.5 POLYSACCHARIDES

Monosaccharides are linked to other sugars through glycosidic linkages to form polysaccharides (carbohydrates). Glycosides can be hydrolyzed to simple building blocks by rupturing the glycosidic bonds. One important class of polysaccharide is represented by cellulose. Cellulose has the formula $(C_6H_{10}O_5)_n$, where n ranges from 500 to 5000, depending on the source of the polymer. Over half of the total organic carbon in the earth's biosphere is in cellulose, for example, cotton and fibers. Wood of bushes and trees contains roughly 50% cellulose.

5.6 GALACTOMANNAN

An important group of polysaccharides is represented by the galactomannans. These polymers contain the monosaccharides mannose and galactose. These monomers are 1-4-β-D-mannopyranose linked and they also have branch points from their six positions linked to α-D-galactose. A representative structure indicating the linkages is shown in Figure 5.7. Galactomannans are found in guar gum where the mannose:galactose ratio is approximately 2:1. They are often added into food

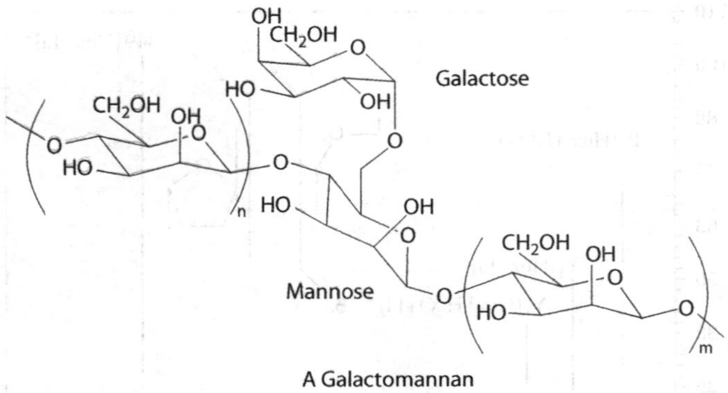

FIGURE 5.7 Example of structure of a polysaccharide galactomannan, indicating both 1-4 or 1-6 linkages.

products to increase the viscosity and used as stabilizers, for example, in ice cream to improve its texture.

Guar gum is an important industrial sugar used as an essential ingredient for mining oil and natural gas in a process called hydraulic fracturing. In recent times, the demand for guar has increased to such a degree that farmers living in poverty in Northwestern India have reaped a windfall and their lives have been transformed due to their ability to harvest the guar gum in their area.

5.7   IMMUNO PROPERTIES OF POLYSACCHARIDES: ECHINACEA

Polysaccharides are purportedly responsible for immune boosting properties. For example, this explanation has been offered for the health benefits of a traditional Western botanical medicine from *Echinacea purpurea* (Figure 5.8). This plant has now been extensively studied and evidence suggests that the polysaccharides contained in echinacea plants possess immune boosting properties.

5.8   POLYSACCHARIDES IN MUSHROOMS: GANODERMA

There is a lot of interest in the health properties of medicinal mushrooms, used for centuries in Asian communities. Well-known medicinal mushrooms include the following species: *Cordyceps* sp., *Grifola* sp., *Lentinus* sp., *Tremella* sp., and *Pleurotus* sp. One particular example of a traditional Chinese medicinal mushroom is *Ganoderma lucidum* (Figure 5.9), also known in China as "lingzhi" or "reishii." Research continues to focus on the immune boosting and antitumor properties of the polysaccharides from these mushroom species.

FIGURE 5.8 Flowers of *Echinacea purpurea*, used in traditional medicine for its immune boosting properties, believed to be due in part to the presence of polysaccharides. (Courtesy of Steven Foster.)

5.9 A NOTE ON THE SEPARATION OF SUGARS

Due to the large number of possible isomers and, in general, the lack of any chromophore, it is difficult to separate and isolate the simple sugars. The separation of polysaccharides is even more challenging. Furthermore, almost all sugars are sensitive to temperature, which not only influences their physical properties but also their chemical stability. It may be fair to say that the isolation and structure determination of sugars continues to be a major challenge, yet to be overcome in natural products chemistry.

The isolation of polysaccharide mixtures from mushrooms can be represented by the example shown below. After the first step to obtain the crude extract, further purification is achieved by the use of gel chromatography. The use of electrochemical detectors (ECDs) in place of UV detectors enhances the detection of eluting polysaccharides in the purification steps. Also, size exclusion gel chromatography can be used to separate polysaccharide fractions with aqueous solvents as eluent.

FIGURE 5.9 The mushroom *Ganoderma lucidum* used in Traditional Chinese Medicine, due in part to the presence of polysaccharides.

Dry, Ground-Up Mushroom Powder

Extract with boiling water
   Centrifuge
   Retain liquid

Pass through membrane filter
   Concentrate liquid
   Advanced chromatography

 ICS-2500 High-Performance Ion Chromatograph (Dionex)

 ED50 Electro Chemical Detector

 Dilute NaOH as eluent

 Complete separation of all key monosaccharides

(Adapted from Zhou, S. et al., 2012, *Intl J Med Mushrooms* 14(4):411–417.)

5.10 FATS AND LIPIDS

Lipids constitute a group of naturally occurring molecules. They represent a large number of important everyday products (e.g., fats, soaps, and waxes); mono-, di-, and triglycerides; phospholipids; and fat-soluble vitamins A, D, E, and K (see Chapter 12). The main biological functions of lipids include energy storage, signaling, and acting as structural components of cell membranes.

5.11 SOAPS

Industrially, soaps are made from a mixture of the Na salts formed by adding NaOH or sodium carbonate to natural fatty acids. The general reaction, called saponification, requires heating up fats and oils with liquid alkali as represented in the following reaction.

$$\text{Fat} + \text{NaOH} \rightarrow \text{Soap} + \text{Glycerol}$$

5.12 DETERGENTS

Detergents are generally used as cleaning materials. They contain one or more surfactants, which reduce the surface tension of water. The detergent consists of a long chain hydrocarbon and an ionic functional group.

- If the ion is an alkyl sulfate, the molecule is anionic.
- If the ionic group is cationic (e.g., an NH_4 group), these detergents are used as a germicide and in shampoos.
- Neutral detergents may include the natural bile salts, which are produced in the liver. Their main function, as derivatives of cholesterol in the body, is to help the digestive process. The bile salts emulsify fats and oils in order that the body's enzymes can further break down the fats.

5.13 WAXES

Waxes are water-resistant and form protective coatings on plants, fruits, and animal skins. Examples of waxlike materials include beeswax; Chinese wax used as a polish; ear wax as a protective layer over the ear membrane; and lanolin found on the skin and fur of animals, and used for rust prevention and cosmetics. Shellac is used as a wood sealant. They differ from fats in that they have no triglyceride esters (of three fatty acids). In fact, they are long chain nonpolar lipid esters, as shown in Figure 5.10, and produced from an alcohol and a long chain fatty acid.

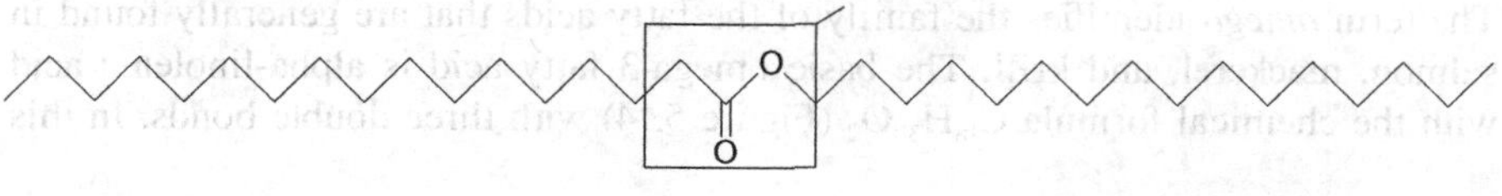

FIGURE 5.10   Structure of a wax ester formed from a long chain fatty acid and an alcohol.

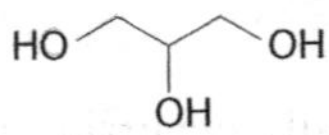

HO⌒⌒OH
OH

FIGURE 5.11   Chemical structure of glycerol.

5.14 TRIGLYCERIDES

In the body, fatty acids are stored in the form of a triglyceride and found in adipose tissue. Once fatty acids are converted into the triglyceride they can enter the bloodstream. These lipid triglycerides can be in the form of fat (solid) or oil (liquid). In general, they are composed of 3 parts fatty acids + 1 part glycerol.

5.15 GLYCOLS

Glycerol has the chemical formula $C_3H_3(OH)_3$ (Figure 5.11).

5.16 FATTY ACIDS

The fatty acids are a major contributor to lipid-like molecules and their chemistry. The naturally occurring fatty acids may be either saturated or unsaturated; the former having higher melting points than the latter of corresponding size. The fatty acids are precursors to numerous compounds, and examples of various structures are illustrated in Figure 5.12.

The essential fatty acids (Figure 5.13) are important to overall human health, yet they cannot be synthesized by the body, therefore they are obtained from the diet. Two important fatty acids, namely linoleic and linolenic, are designated "essential," because their absence in the human diet has been associated with health problems, such as scaly skin, stunted growth, and increased dehydration. A lack of either of these two fatty acids can lead to ill health and causes deficiency symptoms.

Linoleic acid is an integral part of the cell component and is used to manufacture signaling molecules in the body. Linoleic acid is an omega-6 fatty acid. It consists of an 18-carbon chain, a carboxylic acid group, and two *cis* double bonds at positions C9 and C12.

Linolenic acid is a component of cell membranes and used by the body to make substances called eicosanoids, which regulate inflammation. Linolenic acid is an omega-3 fatty acid. It consists also of an 18-carbon chain and a carboxylic acid but possesses three *cis* double bonds. Linolenic acid exists in both the alpha and gamma forms signifying the difference in the location along the chain of the three double bonds.

5.17 SOURCES OF OMEGA OILS

The term *omega* identifies the family of the fatty acids that are generally found in salmon, mackerel, and krill. The basic omega-3 fatty acid is alpha-linolenic acid with the chemical formula $C_{18}H_{30}O_2$ (Figure 5.14) with three double bonds. In this

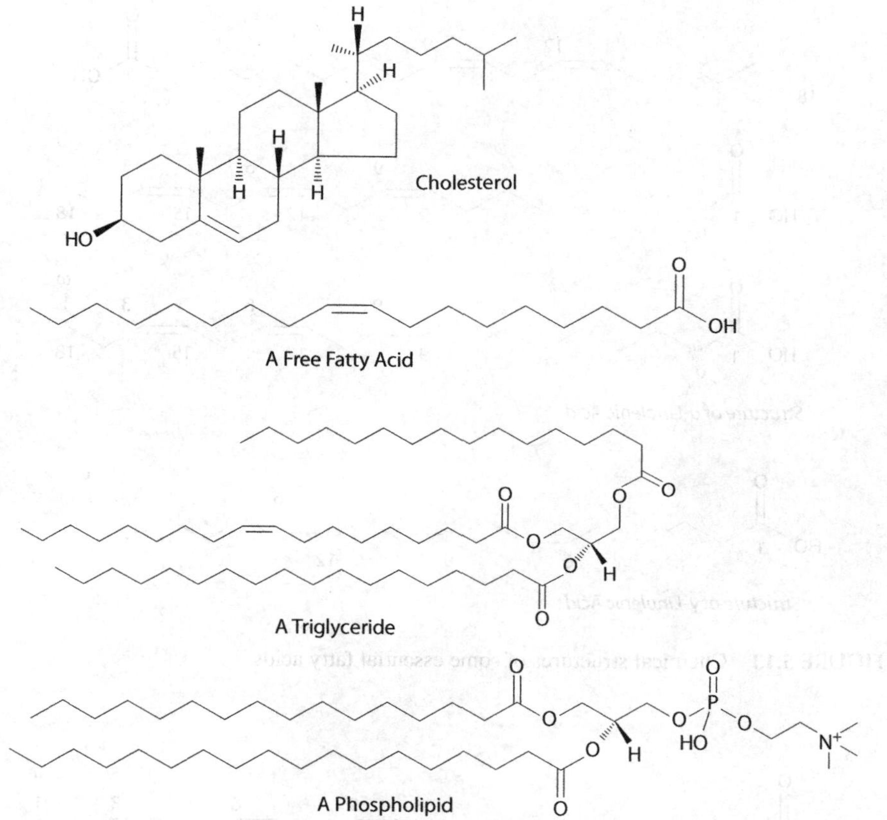

FIGURE 5.12 Examples of various fatty acid compounds found in nature.

nomenclature, the omega-3 fatty acid means it has the first *cis*-double bond, three carbons from the end of its chain.

The polyunsaturated omega-3 oils are found in flax and canola oils, and provide alpha-linolenic acid. These fatty acids offer health benefits in support of endothelial functions and promote synthesis of the vasodilator NO. They are known to reduce platelet aggregation, triglycerides, and blood cholesterol levels.

The polyunsaturated omega-6 compounds can be found in safflower, corn, and sunflower oils; and thus are a good source of linoleic acid. Terrestrial animal oils provide arachidonic acid.

5.18 OMEGA-3 FATTY ACIDS IN FISH OIL

Marine sources provide two important omega oils known as eicosapentenoic acid (EPA) and docosahexenoic acid (DHA) in a ratio 18% EPA and 12% DHA.

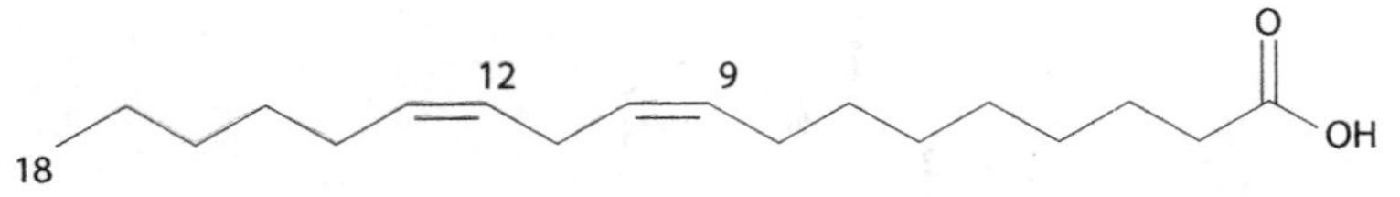

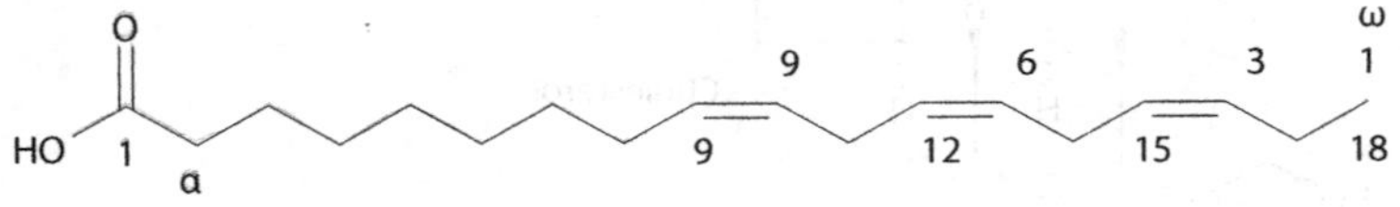

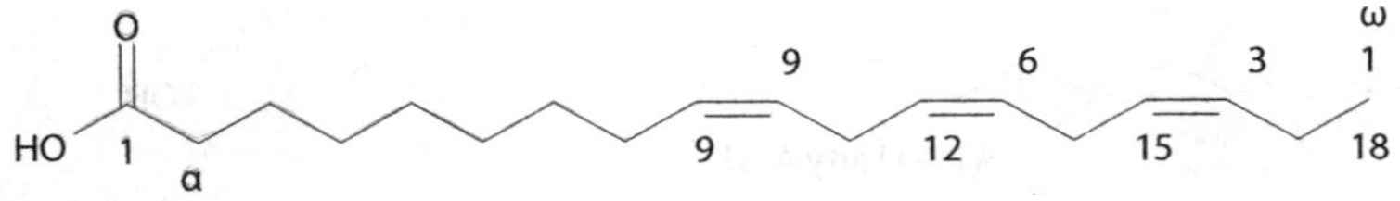

Structure of α-Linolenic Acid

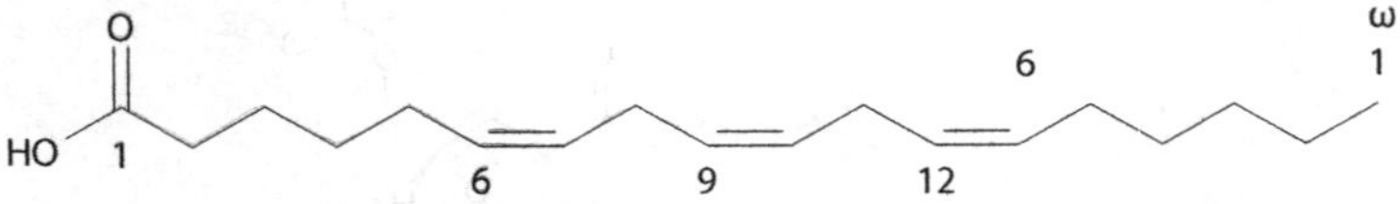

Structure of γ-Linolenic Acid

FIGURE 5.13 Chemical structures of some essential fatty acids.

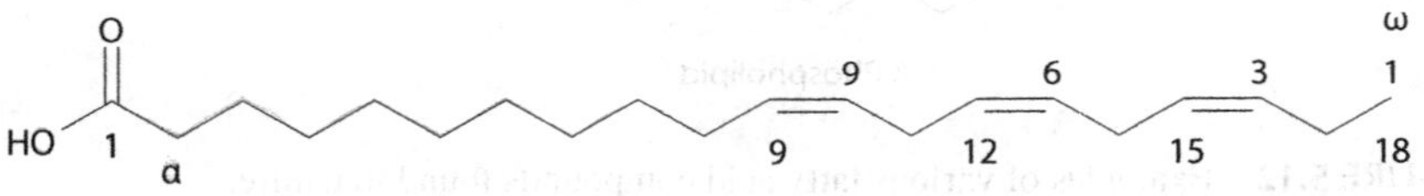

FIGURE 5.14 Chemical structure example of an omega fatty acid.

5.19 SEPARATIONS

The separation of fatty acids has been achieved successfully using gas chromatography (GC) (see Chapter 2, Figure 2.2). Successful GC analysis is best achieved by first converting the fatty acids into their respective methyl esters, which reduces their polarity, and under GC conditions, base line separation is achieved.

5.20 PROSTAGLANDINS, THROMBOXANES, AND LEUKOTRIENES

One very important group of compounds derived from fatty acids is the prostaglandins (Figure 5.15). The prostaglandin contains 20 carbon atoms, including a 5-carbon ring. The prostaglandins, together with the thromboxanes and prostacyclins, are known as the prostanoid class of fatty acid derivatives, which in turn is a subclass of eicosanoids. They are formed in body tissues through the metabolism of the essential

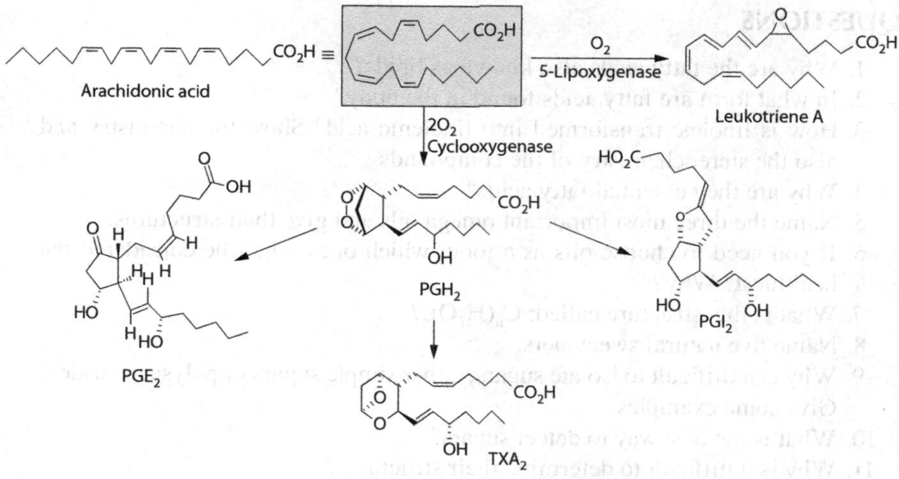

FIGURE 5.15 The prostaglandins.

fatty acid arachidonic acid (5, 8, and 11, 14-eicosatetraenoic acid) to generate eicosanoids. The metabolic pathway by which arachidonic acid is converted to the various eicosanoids is complex.

These compounds constitute a very important family of physiologically potent lipids and are present in minute amounts in most body tissues and possess important functions in the body. Specifically, the prostaglandins are natural hormones with very important biological effects. They lower gastric secretions, stimulate uterine contractions, lower blood pressure, influence blood clotting, and induce asthma-like allergic responses. They also act as mediators and exhibit a variety of physiological effects, such as regulating the contraction and relaxation of smooth muscle tissue.

SUMMARY

In this chapter, two important groups of natural products are presented: the sugars (carbohydrates) and fats (lipids). The carbohydrates include simple sugars, and mono- and polysaccharides. The lipids include the fatty acids, the omega oils, and the prostaglandins.

HISTORICAL NOTE

Prostaglandin derives its name from the prostate gland. It was first isolated in 1935 from prostate secretions. The first total syntheses of prostaglandin $F_{2\alpha}$ and prostaglandin E_2 were achieved in 1969. In 1971, it was determined that aspirin-like drugs could inhibit the synthesis of prostaglandins, and this research work by Samuelsson and Vane was recognized in the award of the Nobel Prize in 1982 for physiology and medicine.

QUESTIONS

1. Why are the fatty acids also known as lipids?
2. In what form are fatty acids found in the body?
3. How is linoleic transformed into linolenic acid? Show the chemistry and also the stereochemistry of the compounds.
4. Why are they essential fatty acids?
5. Name the three most important omega oils and give their structures.
6. If you need to choose oils as a food, which ones might be considered the healthiest? Why?
7. What is this structure called: $C_n(H_2O)_n$?
8. Name five natural sweeteners.
9. Why is it difficult to isolate sugars, either simple sugars or polysaccharides? Give some examples.
10. What is the best way to detect sugars?
11. Why is it difficult to determine their structures?
12. What is the best way to detect sugars?

FURTHER READING

M. G. Enig. 2000. *Know Your Fats: The Complete Primer for Understanding the Nutrition of Fats, Oils and Cholesterol*. Bethesda Pr.
S. Zhou, Q. Tang, X. Luo, J.-J. Xue, Y. Liu, Y. Yang, Ji. Zhang, and N. Feng. 2012. Determination of carbohydrates by high performance anion chromatography-pulsed amperometric detection in mushrooms. *Intl J Med Mushrooms* 14(4):411–417.

6 Phenolic Compounds

Phenolic compounds are produced in nature by a number of different pathways. Two of the most important pathways to produce phenolic compounds are (1) the shikimic acid pathway outlined in Chapter 4 and the metabolic route in plants, and (2) the malonic acid pathway, of more importance in fungi.

Most of the phenolic compounds are derived from the amino acid phenylalanine. Indeed, phenylalanine can be considered the branch point between primary and secondary metabolites. In the steps of the shikimic acid pathway, the removal of the amine function by the enzyme phenylalanine ammonia lyase (PAL) produces the phenylpropanoid compound cinnamic acid. Thus, the two amino acids phenylalanine and tyrosine appear as the precursors in the phenylpropanoid biosynthesis. Furthermore, these phenylpropanoids are then used to elaborate the major phenolic classes, including the flavonoids, coumarins, tannins, anthraquinones, anthocyanins, lignin, and lignans.

6.1 FLAVONOIDS

Flavonoids are ubiquitous in nature and play various roles. Many of these phenolic compounds have antimicrobial activity and are considered to be antioxidants. An important flavonoid found in red wine is known as resveratrol (Figure 6.1), the compound that may be responsible for explaining the "French paradox," as it is believed to lower the risk of heart attacks by inhibiting platelet aggregation and blood clots. In the case of resveratrol, there is no middle oxygenated ring, which is a typical feature of flavone structures. In fact, flavonoids are classified into different groups based on the degree of oxidation of the 3-C bridge. This classification results in structures belonging to the following: anthocyanins, flavones, flavonols, and isoflavones.

In their natural form, the flavonoids also may exist as their corresponding glycosides. During the extraction process the intact glycoside may be isolated or the glycosidic bond is ruptured due to solvent or hydrolysis conditions and only the intact aglycone is recovered.

6.2 FLAVONES

Flavone compounds (Figure 6.2) are the pigmented compounds found in flowers. They strongly absorb ultraviolet (UV) but not visible light, thus are not visible to the human eye but visible to many insects. For this reason, their role in nature may be as attractants and nectar guides. Flavones are present in the leaves, where they protect against UV damage and they may be involved in nitrogen fixation mechanisms.

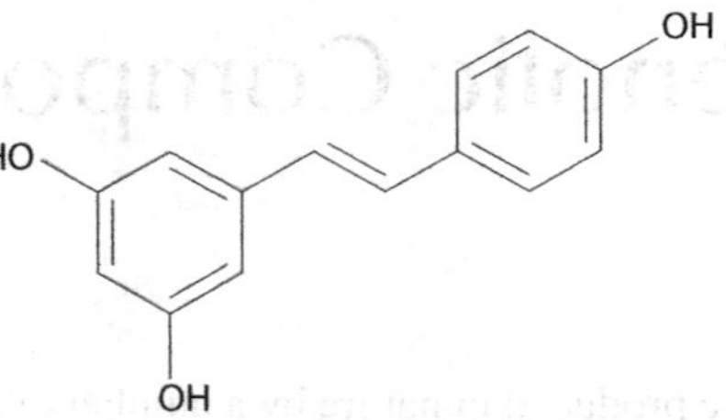

FIGURE 6.1   Chemical structure of resveratrol.

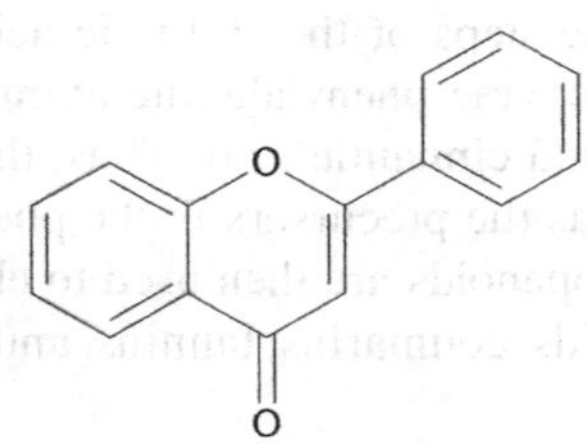

FIGURE 6.2   The flavonoid skeleton.

6.3   ISOFLAVONES

Some isoflavones have strong insecticidal activity (e.g., rotenoids) and others possess estrogenic/antiestrogenic activity, which may cause infertility in mammals. Many are recognized as phytoalexins, antimicrobial compounds.

Several isoflavones are found in the soy plant and they may be responsible for soy's health benefits. The main isoflavone, genistein, purportedly lowers both blood pressure and cholesterol levels, especially LDL, thereby lowering the risk of cardiovascular disease. Genistein (Figure 6.3) has been extensively studied, particularly for its estrogenic effects, which may reduce symptoms of menopause and reduce risks of osteoporosis. Genistein may also inhibit the growth of tumors and is reported to be responsible for low rates of breast and prostate cancer seen among people eating soy products as a major part of their diet.

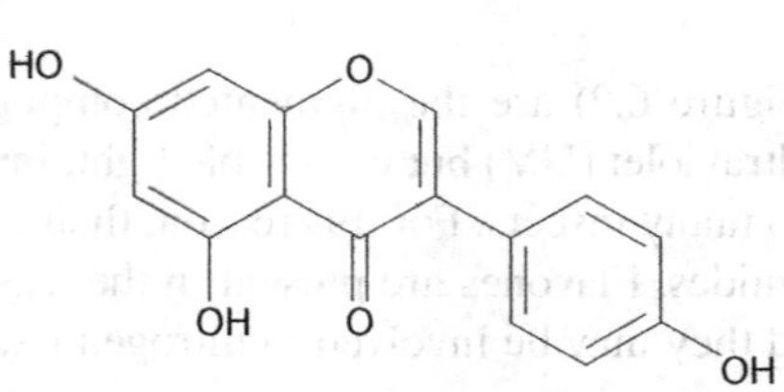

FIGURE 6.3   Chemical structure of genistein, an isoflavonoid found in soy.

6.4 FLAVONE GLYCOSIDES

There are numerous examples in nature of flavonoid glycosides. As an example, in Traditional Chinese Medicine, three major flavone glycosides have been isolated from the leaves of the ginkgo tree, *Ginkgo biloba,* and these compounds are shown in Figure 6.4. Upon extraction from the leaves, three flavone glycosides are isolated. Interestingly, although the glycosides are isolated intact, there has been a preference to perform the chemical analysis of these compounds on their respective aglycones by hydrolysis of the isolated glycosides. Thus, the glycosides, upon hydrolysis yield the corresponding aglycones: quercetin, kaempferol, and isorhamnetin in a ratio of 7:6:1 (see Figure 6.5).

It should be noted that ginkgo leaves contain another important biologically active class of compounds but are not related to the flavonoids: a unique set of terpenoids known as the ginkgolides. Ginkgolides are believed to be potent antagonists against the platelet-activating factor and help prevent platelet aggregation and blood clotting.

Active Constituents of Ginkgo Biloba

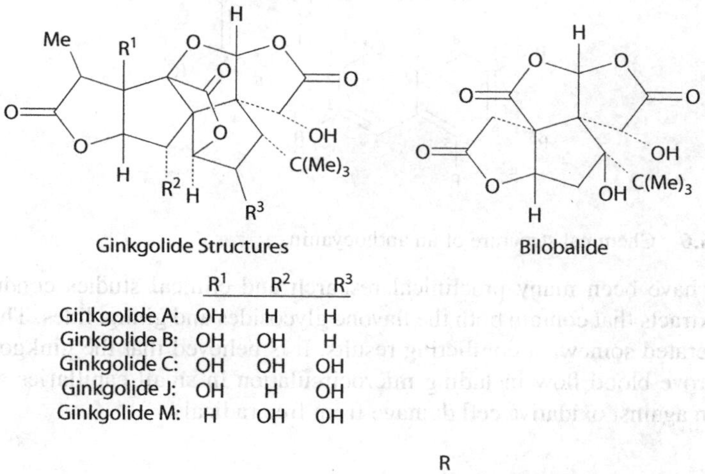

Ginkgolide Structures   Bilobalide

	R^1	R^2	R^3
Ginkgolide A:	OH	H	H
Ginkgolide B:	OH	OH	H
Ginkgolide C:	OH	OH	OH
Ginkgolide J:	OH	H	OH
Ginkgolide M:	H	OH	OH

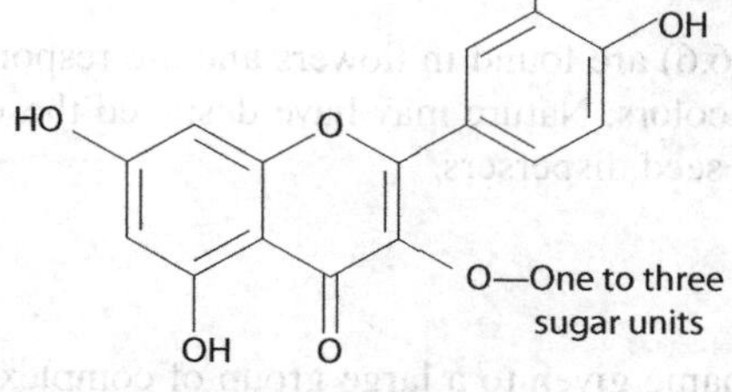

O—One to three sugar units

Flavonol structures
Kaempferol derivatives: R = H
Quercetin derivatives: R = OH
Isorhamnetin derivatives: R = OMe

FIGURE 6.4 Ginkogolide and flavone glycosides found in ginkgo leaves.

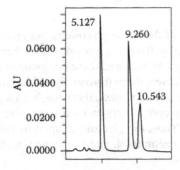

	5.127		
	9.260		
0.0600		Peak Results	
		#	Ret Time (min)	Name

The chromatogram shows peaks labeled 5.127, 9.260, and 10.543 with AU axis values 0.0000, 0.0200, 0.0400, 0.0600.

#	Ret Time (min)	Name
1	5.127	Quercetin
2	9.260	Kaempferol
3	10.543	Isorhamnetin

FIGURE 6.5 HPLC analysis of flavonoid glycosides in a *Ginkgo biloba* extract.

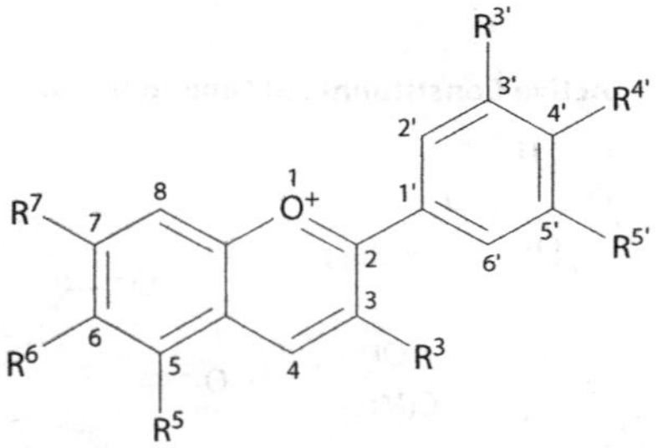

FIGURE 6.6 Chemical structure of an anthocyanin.

There have been many preclinical research and clinical studies conducted on ginkgo extracts that contain both the flavone glycosides and ginkgolides. The studies have generated somewhat conflicting results. It is believed that the ginkgo extracts may improve blood flow including microcirculation in small capillaries and offer protection against oxidative cell damage from free radicals.

6.5   ANTHOCYANINS

Anthocyanins (Figure 6.6) are found in flowers and are responsible for various red, pink, blue, and purple colors. Nature may have designed these molecules to attract animal pollinators and seed dispersers.

6.6   TANNINS

Tannin is the general name given to a large group of complex phenolic substances. The name comes from the leather industry of "tanning" animal hides into leather. Tannins are found in almost every plant part and are particularly abundant in unripe fruit. It is thought they deter herbivores due to their astringent properties and possess antimicrobial activity. A representative structure of tannic acid is shown in Figure 6.7.

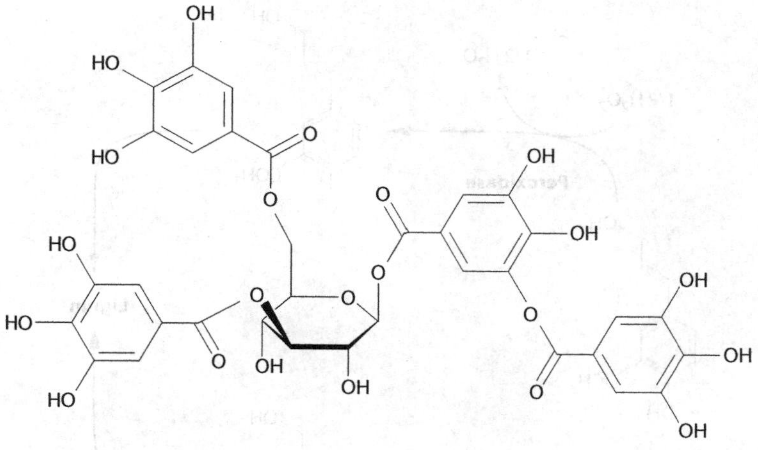

FIGURE 6.7 An example of a tannic acid structure.

6.7 LIGNIN AND LIGNANS

Lignin is a polymer formed by the combination of three phenolic alcohols. These basic monomeric units are the phenyl propanoids (composed of C6-C3 units). The most common units are known as coniferyl, sinapyl, and paracoumaryl alcohols (Figure 6.8). Different plant species have varied ratios of these monomers. For example, lignin from the beech tree has a ratio of 100:70:7 of coniferyl:sinapyl:paracoumaryl alcohols. These phenyl propanoid alcohols readily undergo polymerization (Figure 6.9) through an oxidative and free radical step, and this process is catalyzed by two different oxidative enzymes: peroxidases or oxidases. In this way, the lignin polymeric matrix is created (Figure 6.10).

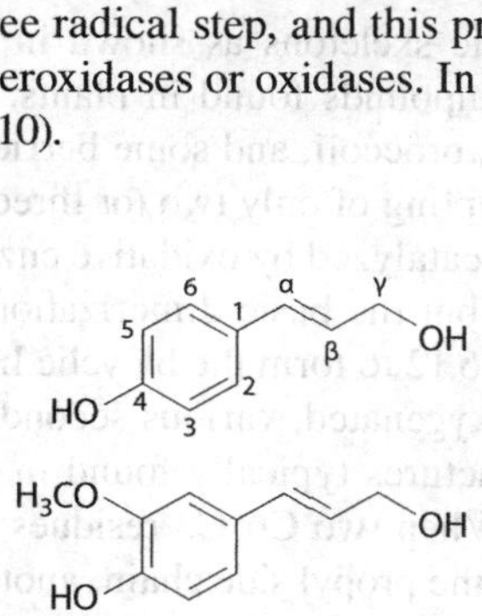

Paracoumaryl Alcohol (1),
Coniferyl Alcohol (2), and Sinapyl Alcohol (3)

FIGURE 6.8 Three common monolignol compounds: precursors to lignin and lignans.

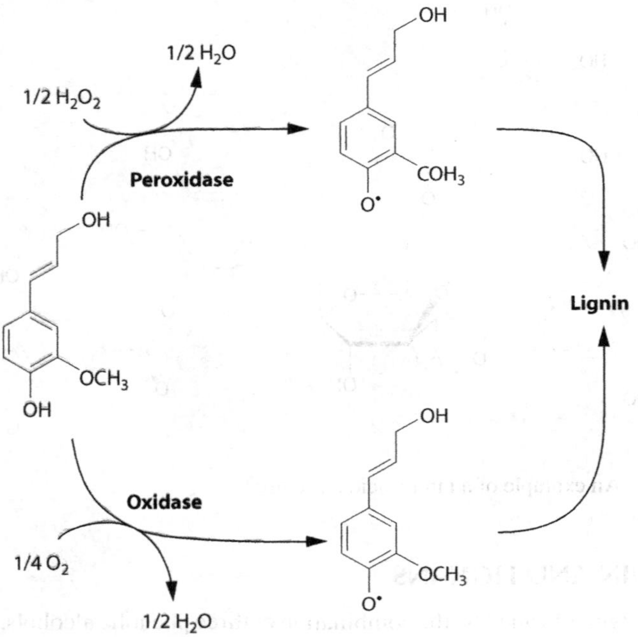

FIGURE 6.9 Polymerization of coniferyl alcohol to a lignin.

6.8 VARIOUS LIGNAN STRUCTURAL UNITS

The lignans form several basic skeletons as shown in Figure 6.11, and represent another group of phenolic compounds found in plants, such as wheat, flax seeds, pumpkin seeds, rye, soybeans, broccoli, and some berries. In contrast to lignin, the lignans are derived via the coupling of only two (or three) substituted C6-C3 monolignols. Again, this reaction is catalyzed by oxidative enzymes. The C6-C3 units are derived from cinnamyl units, but the basic dimerization reaction to form a lignan structure is outlined in Figure 6.12 to form the bicyclic lignan pinoresinol. Since the propyl side chain is usually oxygenated, various secondary cyclizations can occur. Other examples of lignan structures typically found in plants are podophyllotoxin and stegnacin (Figure 6.13). When two C6-C3 residues are linked together but not through the β-carbon atom of the propyl side chain, another class of lignans, known as the neolignans, is formed and represented by skeleton (3) shown in Figure 6.11.

6.9 LIGNANS AS PHYTOESTROGENS

Certain lignans are classified as phytoestrogens similar to the isoflavone genistein. Furthermore, when they form part of the human diet, some lignans are metabolized to form mammalian lignans known as enterodiol and enterolactone by the human intestinal bacteria. Lignans that undergo this transformation include the lignans, matairesinol, and hydroxymatairesinol (Figure 6.14).

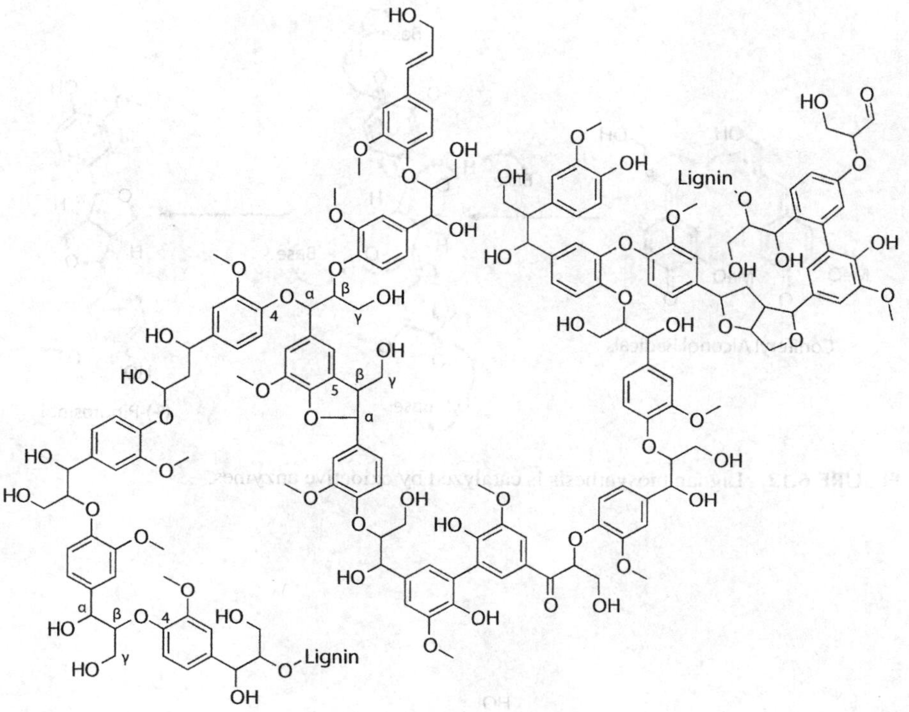

FIGURE 6.10 An example of a lignin polymer backbone.

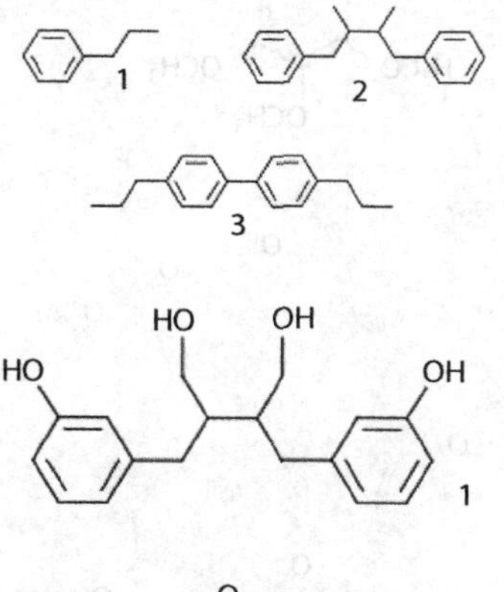

FIGURE 6.11 Examples of various lignan skeletal units.

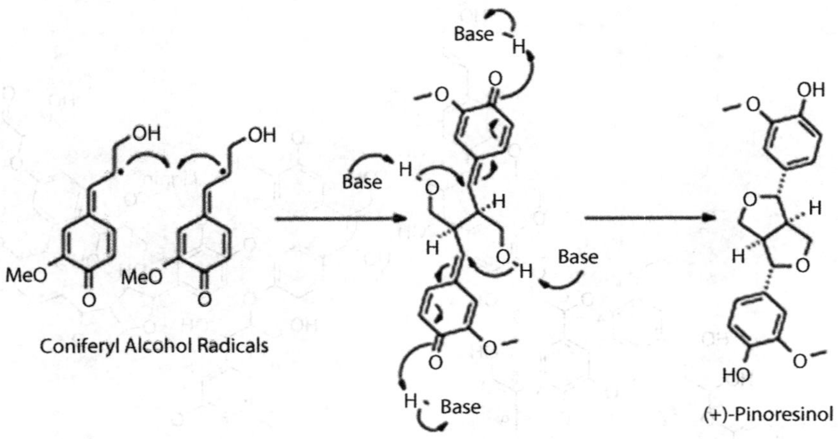

FIGURE 6.12 Lignan biosynthesis is catalyzed by oxidative enzymes.

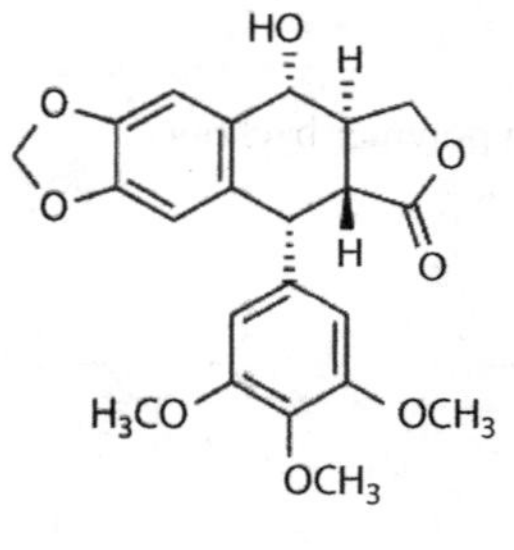

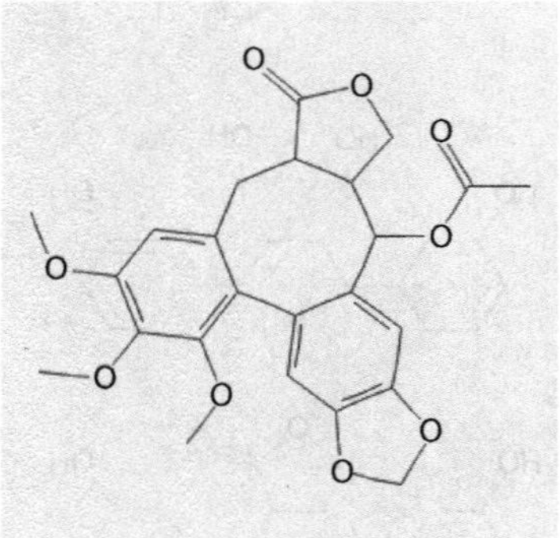

FIGURE 6.13 Other lignan structures represented by podophyllotoxin and stegnacin.

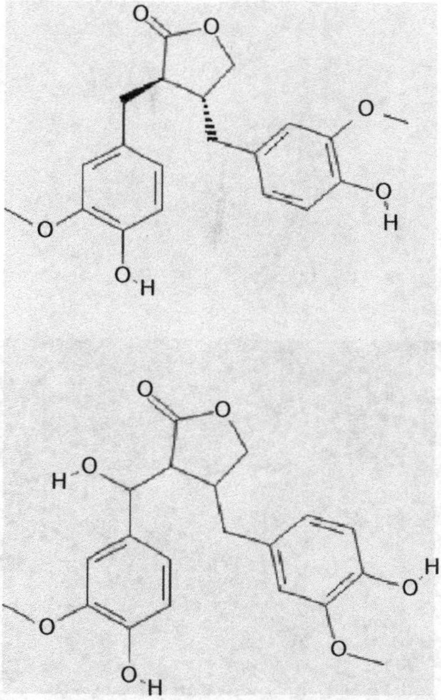

FIGURE 6.14 Structures of the lignans, matairesinol, and 7-hydroxymatairesinol.

6.10 LIGNANS AS GERMINATION INHIBITORS OF WHEAT

Examination of wheat varieties indicate the presence of phenolic compounds including simple phenolic acids, flavonoids, and lignans. For example, in a study examining bran extracts of 16 cereal species, these species were analyzed for lignan content by high-performance liquid chromatography–tandem mass spectrometry. The lignan, 7-hydroxymatairesinol, was the dominant lignan in wheat, together with syringaresinol. Wheat and rye bran had the highest lignan content of all cereals.

It should be noted that the traits that improve wheat as a food source also involve the loss of the plant's natural seed dispersal mechanisms. In fact, the highly

HISTORICAL NOTE

Wheat has always provided an important source of vegetable protein in human food. The grain is easily cultivated, particularly on a large scale, and can be stored after harvest. This source of food enabled settlements to be established at the start of civilization as populations grew in the Babylonian and Assyrian empires, known as the "Fertile Crescent." The three main cash crop cereals in the world today are rice, maize, and wheat (*Triticum* spp.), which accounted for over 680 million tons in 2009.

FIGURE 6.15 Examples of ancient grains. Above: (a) *Triticum boeoticum*, (b) *Triticum dicoccum*, (c) *Aegilops ovata*, and (d) *Hordeum spontaneum*. (Photograph courtesy of Steven Foster.)

domesticated strains of wheat cannot survive in the wild. Some wild species of wheat (the ancient grains; Figure 6.15) can be found growing in the Middle East. A comparison between wild species and their cultivated forms shows a reduction of about 5% of the total phenolic constituents in the latter species. Interestingly, in species of wild wheat there are rare lignans, which act specifically as germination regulators of the wheat seeds. Since these lignans are water-dissolving germination inhibitors they may act as natural "rain gauges." Thus, this aspect may have been especially important for the germination of wheat and similar crops at the appropriate time in those species inhabiting arid regions or deserts, an attribute developed during the domestication of the wild species to the cultivated forms. These structures were determined to be lignans of the type shown in Figure 6.16. Furthermore, in a single oxidative coupling reaction, when one molecule of cinnamoyl alcohol was added to one molecule of the cinnamic acid, as expected three lignans were produced. Compound 3 was identical to the naturally occurring germination inhibitor in the wild wheat (Figure 6.16).

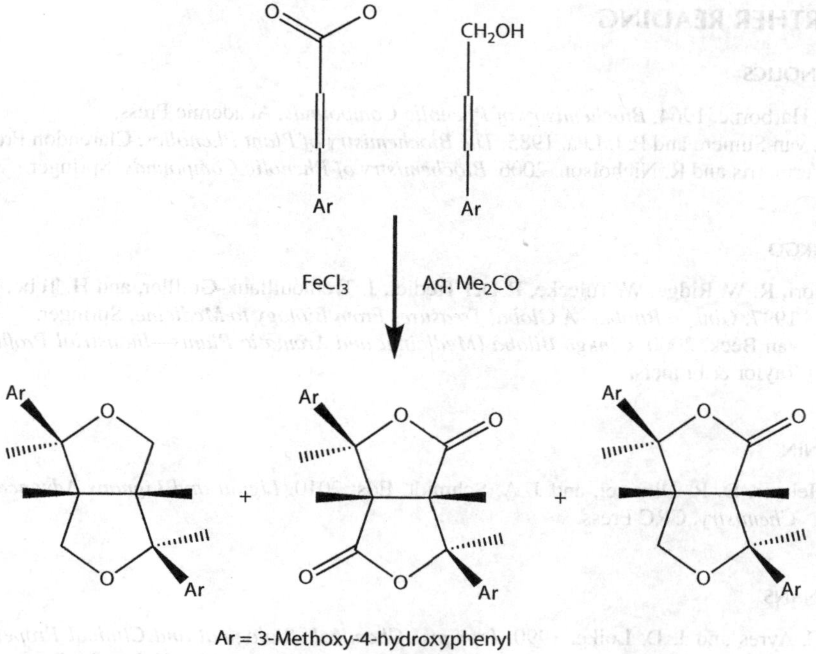

Ar = 3-Methoxy-4-hydroxyphenyl

FIGURE 6.16 Chemistry: Rare lignan, germination inhibitor, found only in select wheat varieties. (From R. Cooper et al., 1994, *J Arid Environ* 27(4):331–336.)

SUMMARY

This chapter introduces phenolic compounds, particularly the examples of flavonoids, flavones and their glycosides, tannins, anthraquinones, coumarins, and anthocyanins. Finally, a description of lignin and lignin compounds is given. These compounds originate from the same building blocks: cinnamic acids, and the C6-C3 phenylpropanoids.

QUESTIONS

1. Ginkgo contains two sets of bioactive compounds. What are they?
2. Draw the structure of bilobalide.
3. What class of compound does it belong?
4. A common flavonoid found in nature is quercetin. What is its structure?
5. What is the key difference between a flavone and isoflavone?
6. The paper industry needs to delignify. Describe the process.

FURTHER READING

PHENOLICS

J. B. Harborne. 1964. *Biochemistry of Phenolic Compounds*. Academic Press.
C. F. van Sumere and P. J. Lea. 1985. *The Biochemistry of Plant Phenolics*. Clarendon Press.
W. Vermerris and R. Nicholson. 2006. *Biochemistry of Phenolic Compounds*. Springer.

GINKGO

T. Hori, R. W. Ridge, W. Tulecke, P. Del Tredici, J. Tremouillaux-Guilller, and H. Tobe, eds. 1997. *Ginkgo Biloba—A Global Treasure: From Biology to Medicine*. Springer.
T. A. Van Beek. 2000. *Ginkgo Biloba (Medicinal and Aromatic Plants—Industrial Profiles)*. Taylor & Francis.

LIGNIN

C. Heitner, D. R. Dimmel, and J. A. Schmidt. Eds. 2010. *Lignin and Lignans Advances in Chemistry*. CRC Press.

LIGNANS

D. C. Ayres and J. D. Loike. 1990. *Lignans: Chemical, Biological and Clinical Properties (Chemistry and Pharmacology of Natural Products)*. Cambridge University Press.
R. Kumar, O. Silakari, and M. Kaur. 2012. *Lignans as Anticancer Agents*. Lambert Academic Publishing.

GRAINS

R. Cooper, D. Lavie, Y. Gutterman, and M. Evenari. 1994. The distribution of rare phenolic type compounds in wild and cultivated wheats. *J Arid Environ* 27(4):331–336.
J. M. Diamond. 1998. *Guns, Germs and Steel: A Short History of Everybody for the Last 13,000 Years*. Vintage.
A. I. Smeds, P. C. Eklund, R. E. Sjöholm, S. M. Willför, S. Nishibe, T. Deyama, and B. R. Holmbom. 2007. Quantification of a broad spectrum of lignans in cereals, oilseeds, and nuts. *J Agri Food Chem* 55 (4):1337–1346.
K. Tanno and G. Willcox. 2006. How fast was wild wheat domesticated? *Science* 311(5769):1886.
D. Zohary and M. Hopf. 2000. *Domestication of Plants in the Old World: The Origin and Spread of Cultivated Plants in West Asia, Europe, and the Nile Valley*. Oxford University Press.

7 Nitrogen-Containing Compounds

7.1 PROTEINS, PEPTIDES, AND AMINO ACIDS

Proteins are present in and vital to all living cells. The protein holds together, protects, and provides structure to the body of multicelled organisms. Proteins catalyze, regulate, and protect the body chemistry in the form of enzymes, hormones, antibodies, and globulins. Furthermore, proteins affect the transport of oxygen and other substances within a living organism. Examples include hemoglobin, myoglobin, and various lipoproteins. Proteins possess positive attributes as antibiotics and vaccines to help fight disease. Conversely, some proteins are recognized as toxins and venoms (see Chapter 14), for example, toxins produced by tetanus and diphtheria microorganisms; snake venoms; and the toxic protein known as ricin, found in castor beans (see Figure 14.4).

7.1.1 COMPONENTS OF PROTEINS

First, it should be noted that proteins are fundamentally different from carbohydrates and lipids. Lipids are made up of approximately 75% to 85% carbon, and carbohydrates contain about 50% oxygen and less than 5% nitrogen. However, the proteins (and peptides) possess between 15% and 25% nitrogen and about an equal amount of oxygen. Other elements include sulfur atoms and are specifically found in the amino acids cysteine and methionine. Proteins and peptides differ in their size: peptides are considered as small proteins, having molecular weights of less than 10,000 daltons. But all proteins and peptides are constructed from amino acids.

7.1.2 PROTEIN–PEPTIDE BUILDING BLOCKS

To date, more than 20 essential amino acids (Figure 7.1) and more than 200 other unusual amino acid compounds have been isolated from nature. The 20 key amino acids are essential diet components. They are not synthesized by human metabolic processes. When proteins are hydrolyzed in boiling aqueous acid or base, these small amino acids are released by rupturing the peptide bonds.

Peptides are synthesized by the coupling of one of the carboxyl group or C-terminus of one amino acid to the amino group or N-terminus of another (Figure 7.2). Peptides can form linear chains of amino acids or can form cyclic peptides of six to eight amino acids or more. Many of the latter are created by microbial and marine sources, and they possess various biological activity.

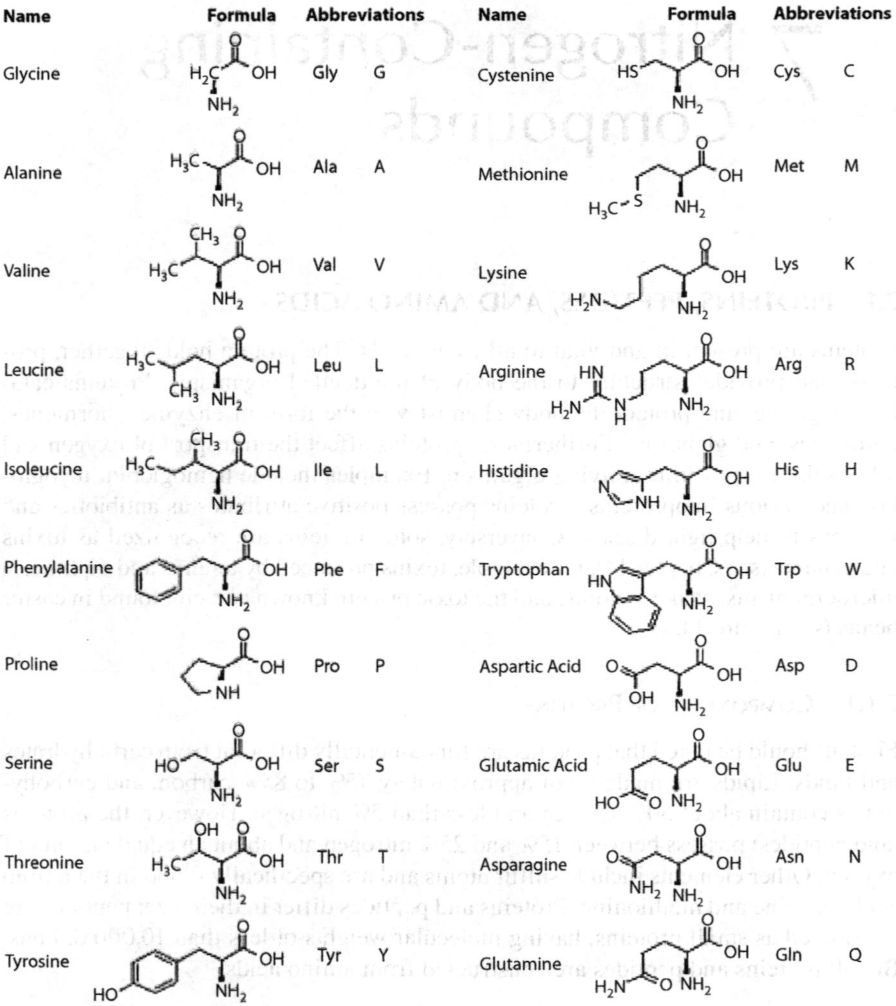

Name	Formula	Abbreviations		Name	Formula	Abbreviations	
Glycine		Gly	G	Cystenine		Cys	C
Alanine		Ala	A	Methionine		Met	M
Valine		Val	V	Lysine		Lys	K
Leucine		Leu	L	Arginine		Arg	R
Isoleucine		Ile	L	Histidine		His	H
Phenylalanine		Phe	F	Tryptophan		Trp	W
Proline		Pro	P	Aspartic Acid		Asp	D
Serine		Ser	S	Glutamic Acid		Glu	E
Threonine		Thr	T	Asparagine		Asn	N
Tyrosine		Tyr	Y	Glutamine		Gln	Q

FIGURE 7.1 List of the common amino acids found in nature.

FIGURE 7.2 Formation of the peptide bond.

7.1.3 Amino Acids

Amino acids can exist in either (1) unionized or (2) a zwitterionic form, whereby a salt is formed when there is a proton transfer from the acidic carboxyl function to the basic amino group to create the zwitterion, represented in Figure 7.3. The isoelectric points of amino acids, in general, range from 5.5 to 6.2. Titration curves showing the neutralization of these acids by added base, and the change in pH during the titration are illustrated in Figure 7.4.

All the natural amino acids listed in Figure 7.1, with the exception of proline, are primary amines. They all possess a stereochemical center except for the amino acid glycine, thus, they are all chiral. Using the Fischer projection, the configurations of the chiral amino acids in nature are the same and defined as the L-configuration by Fischer.

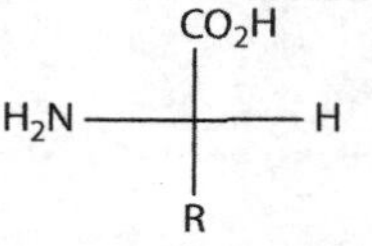

$$H_2N \underset{\underset{R}{|}}{\overset{\overset{CO_2H}{|}}{—}} H$$

L-Amino Acid

$$CH_3CH(NH_2)CO_2H <=> CH_3CH(NH_3)^{(+)}CO_2^{(-)}$$

(1) (2)

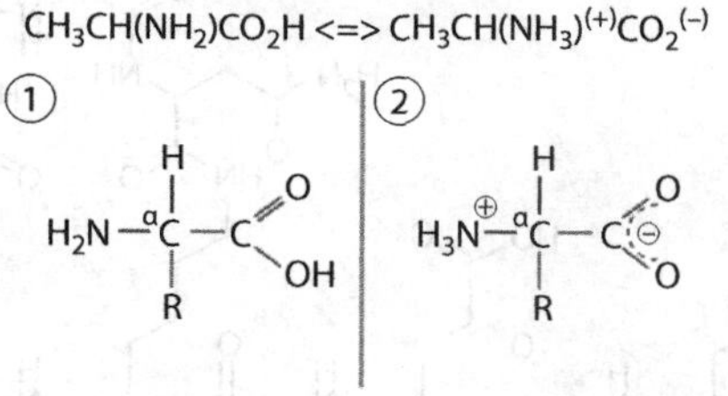

FIGURE 7.3 The zwitterionic forms of amino acids.

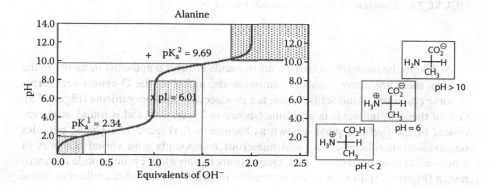

FIGURE 7.4 Graph showing titration curves and the various isoelectric points of amino acids.

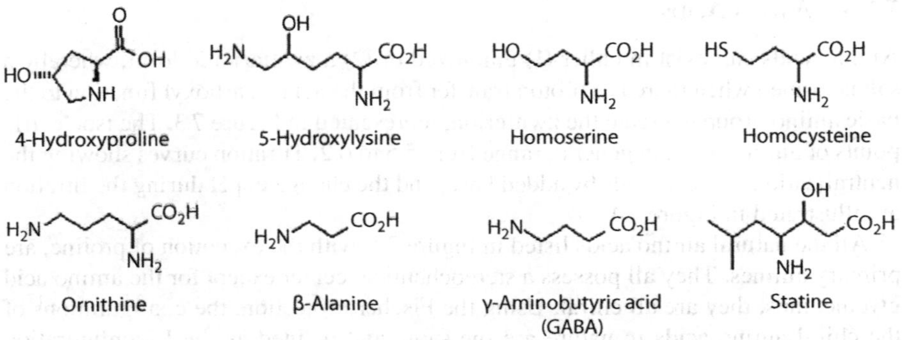

FIGURE 7.5 Examples of other "unusual" amino acids found in nature.

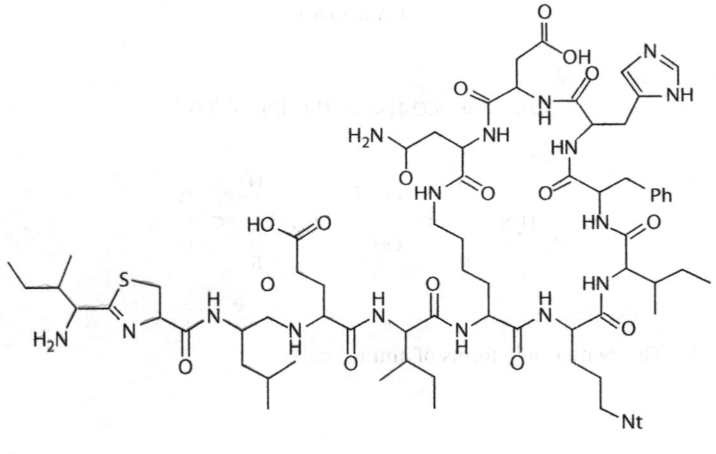

FIGURE 7.6 Structure of the cyclic peptide bacitracin.

Although the naturally occurring amino acids are chiral and exist in nature in the L-form, there are many "unusual" amino acids, including the D-enantiomer forms of some common amino acids, which are produced by microorganisms (Figure 7.5). One of these amino acids is ornithine (shown in Figure 7.5) and is found as a component of the cyclic peptide, known as bacitracin A (Figure 7.6). Cyclic peptides possess antimicrobial activity and numerous compounds with varied numbers of amino acids have been discovered. One extremely important cyclic peptide is vancomycin (Figure 7.7). In view of the increasing resistance to frontline antibiotics, these vancomycins are gaining an ever more prominent role in their use in countering infectious diseases.

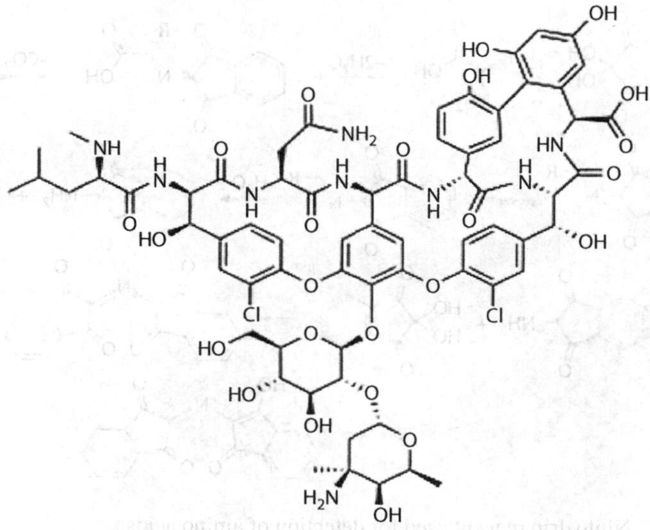

FIGURE 7.7   Structure of the cyclic peptide vancomycin.

7.1.4 DETECTION OF AMINO ACIDS

Many of the amino acids do not possess a chromophore and therefore are not easily visualized by ultraviolet (UV) detection after high-performance liquid chromatography (HPLC) or thin layer chromatography (TLC) separation. Several fluorescent reagents have been used to "tag" amino acids and enhance their detection at extremely low levels. However, one versatile and effective way to detect amino acids and peptides is by the use of ninhydrin (Figure 7.8). The reaction of a free amine with ninhydrin produces a deep blue or purple color when used as a spray reagent to detect amino acids (for example, using TLC, the plate is sprayed with reagent and amino acids appear as purple spots).

7.2 INDOLES AND ALKALOIDS

Indole (Figure 7.9) has a bicyclic structure: a six-membered benzene ring fused to a five-membered nitrogen-containing pyrrole ring. Indole, a major constituent of coaltar obtained by distillation at ~250°C, is used as a common component of fragrances and the precursor to many pharmaceuticals. Synthetic routes have been devised to obtain indoles starting from aniline and ethylene glycol in the presence of catalysts, as shown in Figure 7.10.

One of the more important natural indole compounds, found in the brain, is the amino acid tryptophan, which acts as the precursor of the neurotransmitter serotonin (Figure 7.11). The pathway for the synthesis of serotonin from tryptophan is shown in Figure 7.12.

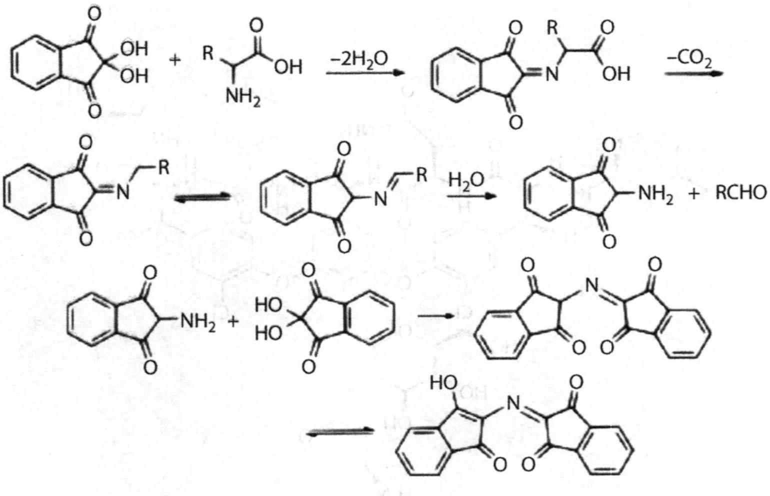

FIGURE 7.8   Ninhydrin reagent used for detection of amino acids.

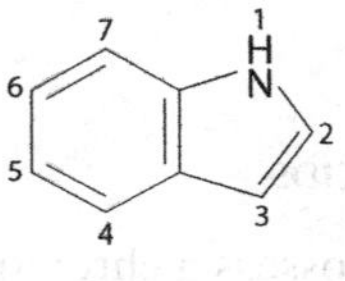

FIGURE 7.9   Basic structure of indole.

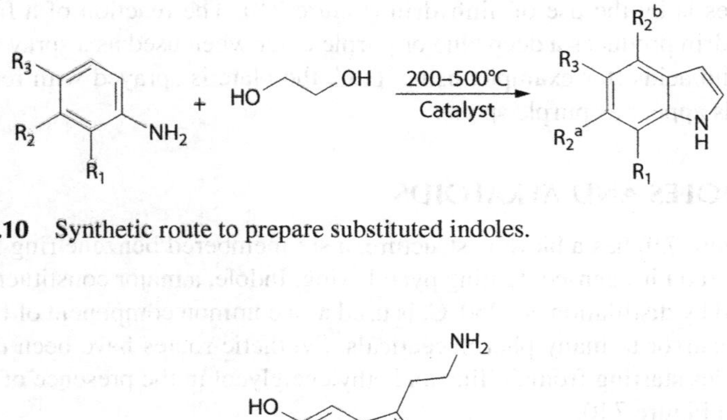

FIGURE 7.10   Synthetic route to prepare substituted indoles.

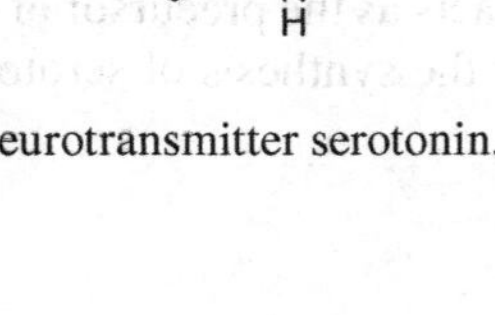

FIGURE 7.11   Structure of the neurotransmitter serotonin.

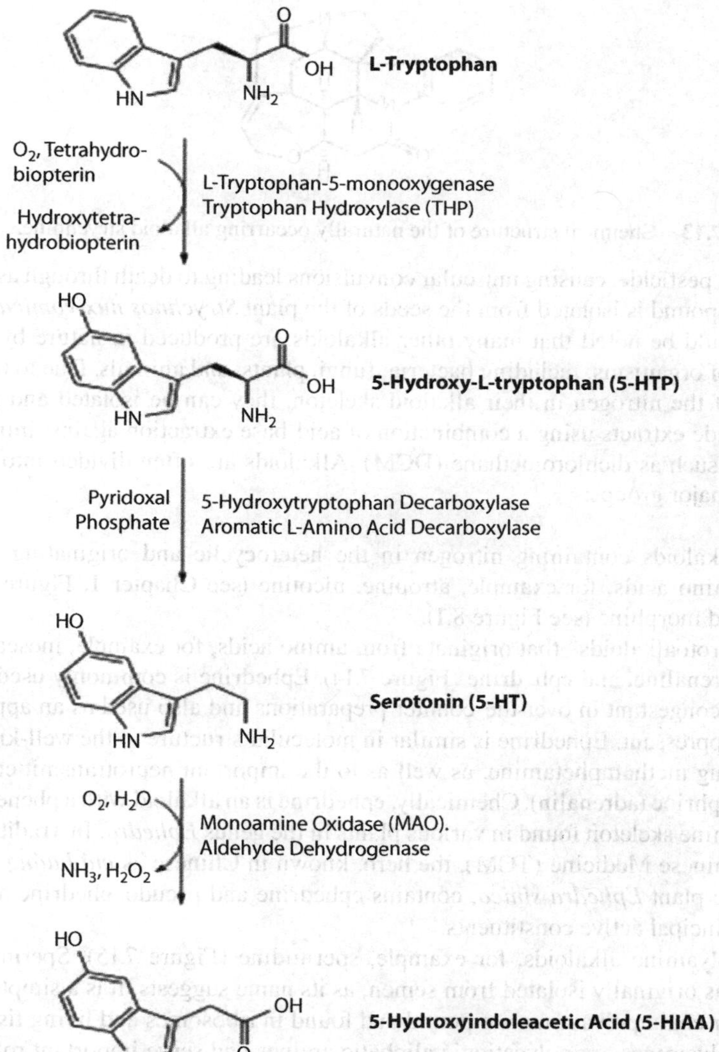

L-Tryptophan

O_2, Tetrahydro-
biopterin

Hydroxytetra-
hydrobiopterin

L-Tryptophan-5-monooxygenase
Tryptophan Hydroxylase (THP)

5-Hydroxy-L-tryptophan (5-HTP)

Pyridoxal
Phosphate

5-Hydroxytryptophan Decarboxylase
Aromatic L-Amino Acid Decarboxylase

Serotonin (5-HT)

O_2, H_2O

Monoamine Oxidase (MAO).
Aldehyde Dehydrogenase

NH_3, H_2O_2

5-Hydroxyindoleacetic Acid (5-HIAA)

FIGURE 7.12 The pathway showing the pathway to serotonin from tryptophan using enzymes and cofactors.

7.3 ALKALOIDS

Naturally occurring alkaloids all contain at least one basic nitrogen atom and usually this atom is located in a cyclic ring system. Alkaloid compounds were among the first natural products to be isolated. Two striking examples include strychnine and morphine (described in detail in Chapter 8) and both compounds originate from plant sources. Strychnine is an indole alkaloid (Figure 7.13) and is highly toxic. It has been

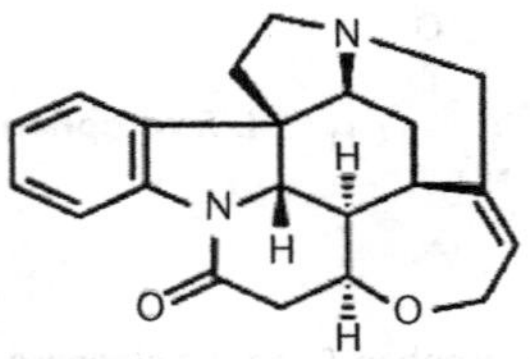

FIGURE 7.13 Chemical structure of the naturally occurring alkaloid strychnine.

used as a pesticide, causing muscular convulsions leading to death through asphyxia. The compound is isolated from the seeds of the plant *Strychnos nux-vomica*.

It should be noted that many other alkaloids are produced in nature by a large variety of organisms, including bacteria, fungi, plants, and animals. Due to the basic nature of the nitrogen in their alkaloid skeleton, they can be isolated and purified from crude extracts using a combination of acid-base extraction against immiscible solvents such as dichloromethane (DCM). Alkaloids are often divided into the following major groups:

1. Alkaloids containing nitrogen in the heterocyclic and originating from amino acids, for example, atropine, nicotine (see Chapter 1, Figure 1.4), and morphine (see Figure 8.1).
2. "Protoalkaloids" that originate from amino acids, for example, mescaline, adrenaline, and ephedrine (Figure 7.14). Ephedrine is commonly used as a decongestant in over-the-counter preparations and also used as an appetite suppressant. Ephedrine is similar in molecular structure to the well-known drug methamphetamine, as well as to the important neurotransmitter epinephrine (adrenalin). Chemically, ephedrine is an alkaloid with a phenethylamine skeleton found in various plants in the genus *Ephedra*. In Traditional Chinese Medicine (TCM), the herb, known in Chinese as *má huáng* from the plant *Ephedra sinica*, contains ephedrine and pseudoephedrine as the principal active constituents.
3. Polyamine alkaloids, for example, spermidine (Figure 7.15). Spermidine was originally isolated from semen, as its name suggests. It is a simple linear chain polyamine compound and found in ribosomes and living tissues. Polyamines are polycationic aliphatic amines and serve important roles in cell survival. Their main function in the body is to synchronize several biological processes (such as Ca^{2+}, Na^+, K^+-ATPase) affecting the membrane potential and controlling intracellular pH and volume.

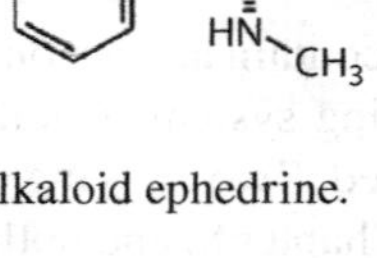

FIGURE 7.14 Example of a protoalkaloid ephedrine.

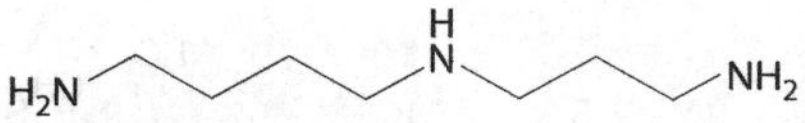

FIGURE 7.15 Chemical structure of a polyamine alkaloid spermidine.

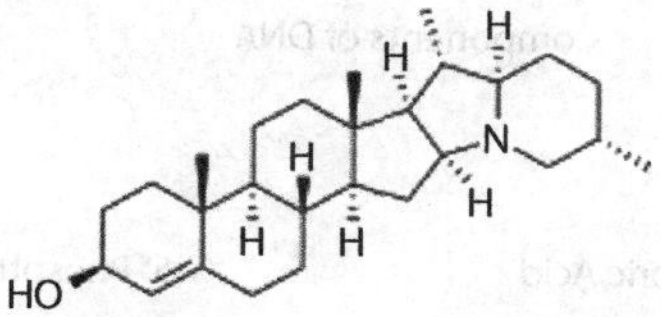

FIGURE 7.16 The steroidal alkaloid solanidine.

4. Peptide and cyclopeptide alkaloids (e.g., Figures 7.6 and 7.7).
5. Pseudoalkaloids, which do not originate from amino acids, for example, terpene-like and steroid-like alkaloids. Solanidine (Figure 7.16) is an example of a steroidal alkaloid. In fact, it is a quite toxic substance, responsible for neuromuscular syndromes via cholinesterase inhibition. The compound or its glycoside is found in plants of the Solanaceae family, such as the potato.
6. Purine-like alkaloids, for example, caffeine (Chapter 2, Figure 2.7), theobromine, and theophylline.

7.4 NUCLEIC ACIDS AND BASE PAIRINGS

Nucleic acids are large biological polymers and include deoxyribonucleic acid, better known as DNA, and ribonucleic acid (RNA). They constitute the most important biological macromolecules; each is found in abundance in all living things. Their function is to encode, transmit, and express genetic information.

Beginning in the 1920s, nucleic acids were found to be major components of chromosomes. These are small, gene-carrying bodies in the nuclei of complex cells. In addition to the elements C, H, N, and O, elemental analysis of nucleic acids revealed the presence of phosphorus. However, unlike some proteins, the nucleic acids do not contain any sulfur. When the nucleic acid is completely hydrolyzed, four different heterocyclic bases are obtained together with inorganic phosphate and the sugar 2-deoxyribose. These individual building blocks are shown in Figure 7.17. The 2-deoxyribose is attached to one of the bases to form a nucleoside. Further, when the phosphate is also attached to the sugar in the form of a 5′-phosphate and the sugar is further connected with a base, this unit is now called a nucleotide. Although the nucleotide consists of three components, the base can be either a purine or a pyrimidine. There are four bases in the DNA molecule and they are known as thymine (T), adenine (A), guanine (G), and cytosine (C). In RNA, the uracil (U) replaces thymine and ribose replaces 2-deoxyribose.

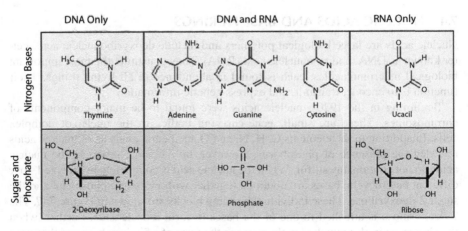

FIGURE 7.17   The building blocks of DNA and RNA.

7.5   RNA AND PROTEIN SYNTHESIS

The genetic information stored in DNA molecules is used as a blueprint for making proteins. These macromolecules have diverse primary, secondary, and tertiary structures that allow them the ability to function and maintain integrity of the living organism.

A dependent DNA polymerase enzyme, known as a reverse transcriptase, is able to transcribe a single-stranded RNA sequence into double-stranded DNA. Such enzymes are found in all cells and are an essential component of retroviruses. These require RNA replication of their genomes. It should be noted that direct translation of DNA information into protein synthesis has not yet been observed in a living organism. Finally, proteins appear to be an informational dead end and do not provide a structural blueprint for either RNA or DNA. Structural information translation is always from mRNA to protein.

SUMMARY

This chapter introduces the nitrogen-containing molecules: proteins, peptides, and amino acids (the building blocks and precursors to linear peptides and some cyclic peptides). The chapter continues with the nucleic acids and their bases, followed by examples of alkaloids. The amino acids tryptophan, phenylalanine, and tyrosine are shown as important precursor building blocks for alkaloids.

QUESTIONS

1. Give chemical structures of the following alkaloids: atropine, nicotine, and morphine.
2. Give the structures of the following:
   a. Caffeine, theobromine, and theophylline
   b. Mescaline, adrenaline, and ephedrine
   c. Putrescine, spermidine, and spermine
   d. One example of a cyclopeptide alkaloid
   e. One example of a terpene-like alkaloid
   f. One example of a steroid-like alkaloid

FURTHER READING

PROTEINS

C. I. Brändén and J. Tooze. 1999. *Introduction to Protein Structure*. Garland Pub.
A. Kessel and N. Ben-Tal. 2012. *Introduction to Proteins: Structure, Function, and Motion*. Taylor & Francis.

ALKALOIDS

E. Fattorusso and O. Taglialatela-Scafati. 2008. *Modern Alkaloids: Structure, Isolation, Synthesis, and Biology*. Wiley.
M. Hesse. 2002. *Alkaloids: Nature's Curse or Blessing?* Wiley.

NUCLEIC ACIDS

V. A. Bloomfield, D. M. Crothers, and I. Tinoco. 2000. *Nucleic Acids: Structures, Properties, and Functions*. University Science Books.
P. O. P. Ts'o. 2012. *Basic Principles in Nucleic Acid Chemistry*. Elsevier Science.

Section III

Natural Product Contributions to Human Health

8 Euphorics

Plants that generate a euphoric effect in humans and possess psychoactivity have played an important role in the culture of mankind and with indigenous and ethnic minorities. There are prescription drugs that are known for their euphoric effects. Examples include morphine and codeine derived from opium. In some countries there are illicit drugs: cocaine from the coca leaf and marijuana from cannabis, all described further in this chapter. There are also euphoric compounds that have a degree of social acceptance, such as nicotine (Figure 1.4) in tobacco and caffeine (Figure 2.7) in tea and coffee.

8.1 MORPHINE

Morphine is a natural alkaloid (Figure 8.1) produced in the poppy plant *Papaver somniferum* (Figure 8.2). This plant usually grows in arid climates, typically in regions of the Far East, such as Iran and Afghanistan. When ripe for harvesting the poppy produces a white sticky latex called opium. Farmers harvest opium 2 weeks after the petals fall from the bud. They make two or three incisions into the pod's skin using a sharp blade, which allows the latex to ooze out slowly to harden over a 24-hour period on the outside of the pod, at which time it is then collected. This latex, or opium, is a complex mixture containing at least 50 different alkaloids. The most abundant alkaloid is morphine, which makes up 8% to 17% of the dry weight of opium.

Although it has been used for centuries, the chemical structure of morphine was only determined in 1925. The vast majority of morphine continues to be harvested from the opium poppy, however, there are at least three ways of synthesizing morphine from simple starting materials such as coal tar and petroleum distillates.

8.2 ISOLATION OF MORPHINE

Morphine and related alkaloids can be purified from opium resin and crude extracts by extraction in the following manner: first, soaking the resin with diluted sulfuric acid, which releases the alkaloids into solution. The alkaloids are then precipitated

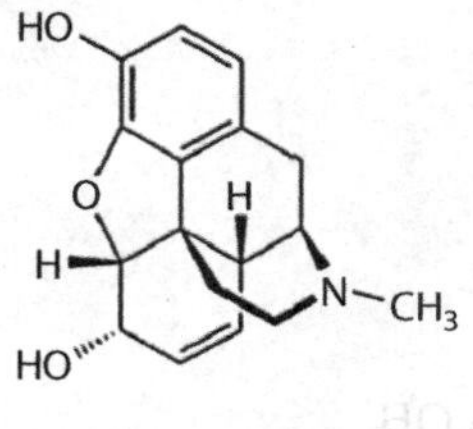

FIGURE 8.1 Chemical structure of the alkaloid morphine.

FIGURE 8.2 The opium-producing poppy *Papaver somniferum*. (Courtesy of Steven Foster.)

by either ammonium hydroxide or sodium carbonate. The last step separates morphine from other opium alkaloids. Today, morphine is isolated from opium in relatively large quantities: over 1000 tons per year.

Raw Opium

   Reflux with MeOH
Filtration

MeOH extract

   Remove MeOH

Total alkaloids

   1.0 M NaOH(aq)
and NaOAc

Filtrate

   Extract with toluene

Retain aqueous phase

   Adjust pH to 9 with AcOH

Morphine precipitate

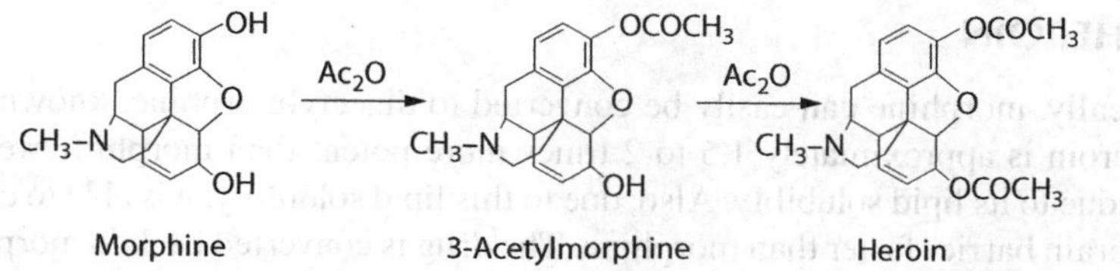

FIGURE 8.3 Acetylation of morphine using acetic anhydride to generate heroin.

8.3 HOW MORPHINE WORKS IN THE BRAIN

Morphine is used as a powerful analgesic to relieve severe pain by acting directly on the brain. It appears to mimic endorphins, natural substances produced by the brain that are responsible for reducing pain but also cause sleepiness and feelings of pleasure in the body. Morphine is classified as a narcotic, which is a drug that dulls the senses. Although it has painkilling properties it also possesses both euphoric and hallucinatory effects.

Morphine also can be chemically converted by an acetylation reaction using acetic anhydride and pyridine to create a much more potent form of narcotic drug known as heroin (Figure 8.3) (see Section 8.5).

8.4 CODEINE

Morphine is converted to another alkaloid with a similar skeleton known as codeine (Figure 8.4) by chemical synthesis using an industrial methylation procedure. Thus, codeine becomes by far the most commonly used opioid in the world. There are several methods for converting codeine to morphine by demethylation (codeine is morphine 3-methyl ether). One way is by the cleavage of aromatic ethers by refluxing morphine with concentrated HBr or HI. To produce codeine in another direct way, albeit with low yield, is to extract codeine from the opium, after separation of the more abundant morphine. Usually, however, most codeine is synthesized from morphine through an O-methylation reaction.

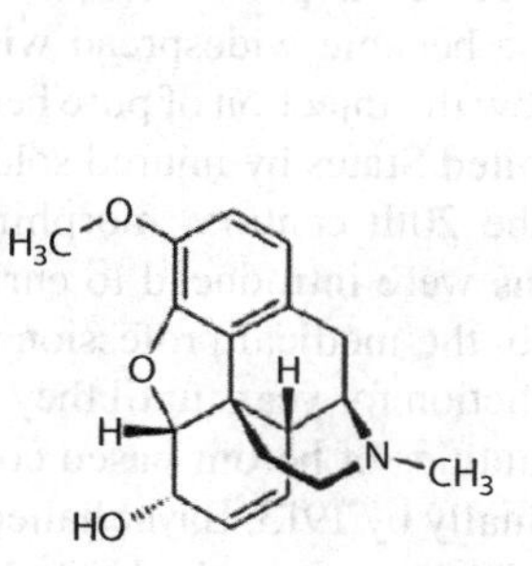

FIGURE 8.4 Chemical structure of a related alkaloid to morphine, codeine, where one of the OH groups undergoes a methylation reaction.

8.5 HEROIN

Chemically, morphine can easily be converted to diacetylmorphine, known as heroin. Heroin is approximately 1.5 to 2 times more potent than morphine weight for weight due to its lipid solubility. Also, due to this lipid solubility, it is able to cross the blood-brain barrier faster than morphine. The drug is converted back to morphine by enzymes in the body before it binds to opioid receptors. These receptors are located in many areas of the brain, and affect the perception of pain and reward. Opioid receptors are found also in the spinal cord and digestive tract. However, heroin is an illicit narcotic drug.

8.6 MEDICAL USES OF MORPHINE

Morphine is used to relieve severe or agonizing pain and suffering by acting directly on the brain. The endorphins are released in response to pain, strenuous exercise, or excitement. In clinical settings, morphine exerts its principal pharmacological effect on the brain. Its primary actions of therapeutic value are to reduce pain and make patients sleepy.

HISTORICAL NOTE

Ancient peoples either ate parts of the poppy flower or converted them into liquids to drink. By the 7th century, the Turkish and Islamic cultures of Western Asia had discovered that the most powerful medicinal effects could be obtained by smoking the poppy's congealed juices and the habit spread. Whereas Indians ordinarily ate opium, the Chinese mixed Indian opium with tobacco and pipe smoking was prevalent throughout the region.

Due to the work of Paracelsus (1490–1541), laudanum ("something to be praised") was created by extracting opium into brandy. By the 19th century, vials of laudanum and raw opium were widely available in England and served as a medicine, not a drug of abuse. The Bayer pharmaceutical company in Germany maintained a large heroin production, as the use of opium for medicine was still legal. Its use became widespread with the development of the hypodermic syringe to allow the injection of pure heroin. There was also heavy use of morphine in the United States by injured soldiers after the Civil War.

At the beginning of the 20th century, morphine addiction became well understood and regulations were introduced to curtail opium as an over-the-counter medicine. Even so, the medical profession remained largely unaware of the potential risk of addiction for years until they noticed their patients were consuming inordinate quantities of heroin-based cough remedies. Opium was not the miracle cure and finally by 1913, Bayer halted production. Finally, morphine became a controlled substance in the United States under the Harrison Narcotics Tax Act of 1914.

HISTORICAL NOTE

Although the stimulant and hunger-suppressant properties of coca had been known for many centuries, the cocaine alkaloid was only first isolated in 1855. The first synthesis and confirmation of the structure occurred in 1898. Coca was first traded as a legal drug in the 19th century but became listed as an illicit narcotic drug in the 1980s.

8.7 CANNABIS

The Cannabis plant has long been used for fiber (hemp) and for medicinal purposes. However, it is more popularly known in the form of a recreational drug. The principal psychoactive constituent responsible for the "high" associated with cannabis, also referred to as marijuana or hashish, is consumed orally or smoked. The active chemical constituent is called phytocannabinoid Δ^9-tetrahydrocannabinol (THC) (Figure 8.5). To date, there are at least 85 different cannabinoids that have been isolated from cannabis. THC acts on the cannabinoid receptors on cells in the body that repress neurotransmitter release in the brain.

8.8 COCAINE

Cocaine (Figure 8.6) is an alkaloid, obtained from the plant *Erythroxylum coca,* found in South American countries such as Peru, Bolivia, and Colombia. After being processed into a paste, the resulting coca base is processed into fine cocaine powder. Cocaine is a stimulant and provides a "mood lift"; the impact to society in general as an illicit drug has been quite significant.

SUMMARY

In this chapter, natural products giving a state of euphoria are presented, offering three important examples: morphine (heroin), the cannabinoids (marijuana), and cocaine.

Morphine, opium, codeine, and heroin are related chemicals known as alkaloids that are considered narcotic drugs and are controlled substances in most countries around the world. Most morphine is derived from the opium poppy and codeine is made from morphine by chemical conversion.

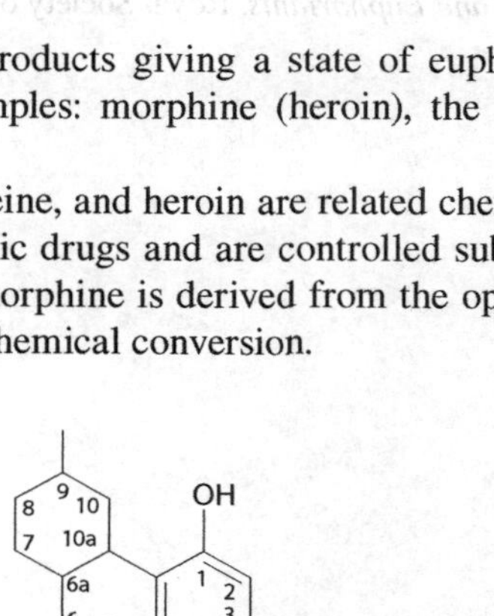

FIGURE 8.5 The active chemical constituent, Δ^9-tetrahydrocannabinol (THC), in *Cannabis.*

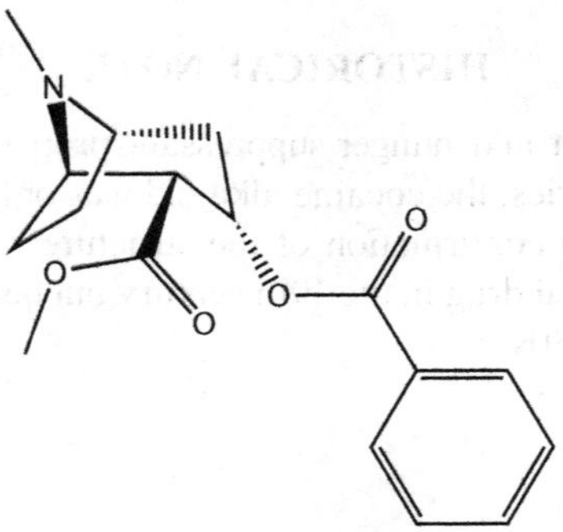

FIGURE 8.6   The alkaloid cocaine.

QUESTIONS

1. What is thebaine?
2. How is the alkaloid thebaine related to morphine and codeine?
3. Give two more examples of cannabinoid structures.
4. What is a good approach to the isolation of cannabinoids from plants?
5. How might you isolate cocaine from coca leaves?
6. Why is acid base extraction helpful to purify alkaloids?

FURTHER READING

R. Cooper. 2013. Morphine and heroin. The yin and yang of narcotics. *Chem Matters* Dec:14–16.

T. Dalrymple. 2008. *Romancing Opiates: Pharmacological Lies and the Addiction Bureaucracy.* Encounter Books.

J. Mann, J. Emsley, P. Ball, P. Page, J. P. Michael, and H. Oakeley. 2009. *Turn on and Tune in: Psychedelics, Narcotics and Euphoriants.* Royal Society of Chemistry.

9 Anti-Infectives from Nature

9.1 ANTIMICROBIAL β-LACTAMS

There are many ß-lactam antibiotics. They are characterized by a four-membered ß-lactam ring (Figure 9.1). Examples include penicillins, cephalosporins, clavams (or oxapenams), cephamycins and carbapenems, monobactams, and nocardicins; almost all derived initially from microbes through a fermentation process.

Penicillin G (Figure 9.2) was the first β-lactam antibiotic that entered the clinic. It has a bicyclic structure consisting of a four-member β-lactam ring and a five-member hydrothiazole ring. The activity of these antibiotics is related to the opening of the ß-lactam ring. A large ring strain leads to both its high antibacterial potency and its instability under acidic and basic conditions.

HISTORICAL NOTE

In 1928, the UK microbiologist Alexander Fleming made the "accidental" discovery that a metabolite produced by the strain of blue-green mold could resist the *Staphylococcus aureus* bacterium. He did not pursue this work further. In 1938, Howard Florey and Ernst Chain began work on the extraction of penicillin from fermented broth, followed in the 1940s by purification of large amounts of penicillin from a strain of *Penicillium notatum*. However, penicillin is acid labile and resulted in low yields. The scientist Norman Heatley developed a back extraction method by controlling the pH of the broth and using a higher producing strain, *Penicillium chrysogenum*. In this manner over 200 times more penicillin was produced by *P. chrysogenum* than *P. notatum*.

The power of coupling microbiology, strain development, and fermentation technology took on an altogether major step during World War II (1939–1945). The significance and importance was not lost on the allied governments, as large numbers of soldiers were dying from secondary infections in combat. Thus, the demand for a large amount of penicillin to be produced saved the lives of soldiers who otherwise would have died due to infection of their wounds. Yet remarkably, although prescribed to humans, the structure and purity of penicillin was not fully determined until after the war, a situation that today would not be acceptable to regulatory bodies such as the U.S. Food and Drug Administration (FDA).

Today, modern forms of penicillin are the frontline antibiotics to counter infections, caused by gram positive and gram negative bacteria, and attack diseases such as pneumonia and gonorrhea.

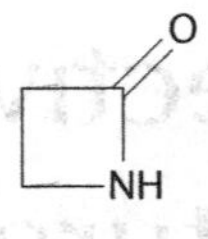

FIGURE 9.1   The basic beta-lactam ring structure.

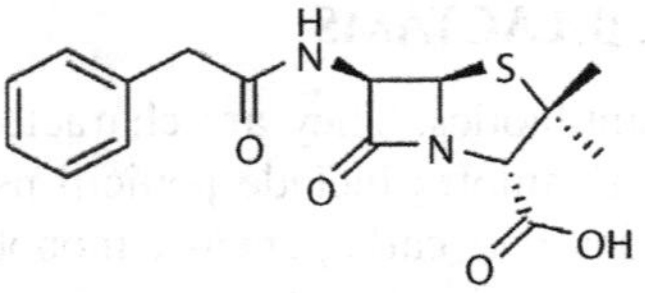

FIGURE 9.2   Chemical structure of penicillin G.

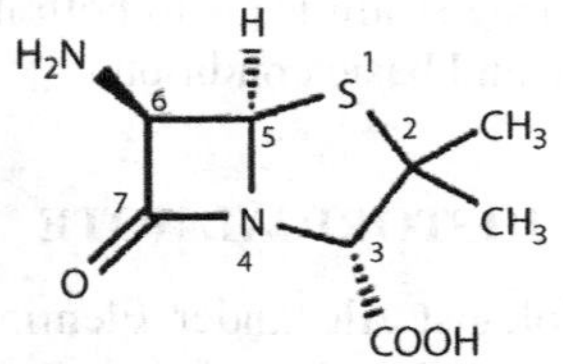

FIGURE 9.3   The structure of 6-aminopenicillanic acid, also known as 6-APA.

There were limits to penicillin G's clinical application, however, due to (1) the presence of bacteria resistant to penicillin G, (2) the route of needing parenteral administration, and (3) allergic susceptibility.

In 1957, the fundamental structure unit of penicillin G, 6-aminopenicillanic acid (6-APA) (Figure 9.3), was announced. This discovery led to the production of new semisynthetic penicillins. Chemical modifications were achieved on the 6β-amino, the 6α, and the C3-carboxylic acid groups and these modifications significantly improved the stability and potency. By introducing a moderately and sterically hindered substitution group at the 6α position, this chemistry enhanced the resistance to β-lactamase, an enzyme produced by bacteria to destroy penicillin. Currently, only penicillin G and penicillin V are naturally occurring approved drugs, while the others are all semisynthetic products.

9.2   STRUCTURE ELUCIDATION OF PENICILLIN

There are three chiral carbon centers in the 6-APA molecule. However, only the isomer with the absolute configuration of 3S, 5R, and 6R possesses bioactivity. A characteristic spectroscopic feature of β-lactams is featured in the infrared (IR) spectrum (e.g., 6-APA is shown in Figure 9.4). The v_{max} (cm^{-1}) of 1780 and 1660 are diagnostic IR stretching bands of the β-lactam: representing C=O and amide C=O

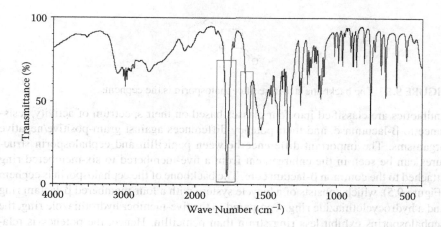

FIGURE 9.4 IR spectrum of penicillin, 6-APA, clearly indicating the strong lactam C=O stretch around 1780 cm^{-1}.

stretches, respectively. Further key structural features of penicillin compounds and confirmation of their structures are supported by additional nuclear magnetic resonance (NMR) and mass spectrometry (MS) spectral data.

9.3 ISOLATION OF PENICILLIN

The classical isolation method to obtain penicillin G from fermentation broth is based on liquid–liquid extraction. The disadvantages of this method are (1) large consumption of organic solvent and (2) low recovery rates. An alternate, more recent method is now used to isolate penicillin G from the fermentation broth and this process is known as aqueous two-phase extraction (ATPE).

The mechanism of ATPE is based on an adsorptive bubble separation technique in which the surface-active compounds in water are adsorbed on the bubble surfaces of an ascending gas stream and then collected in an organic phase as a layer placed on top of the water column.

Typically, this process is carried out at room temperature and, first, upon completion of the fermentation, the broth is mixed with ammonium sulfate. This solution is then adjusted to pH 6.8 with small amounts of HCl solution and NaOH solution, and then transferred to a floating ion cell. The penicillin G floats by bubbling nitrogen gas at a flow rate of 40 ml/min from the bottom of the cell for 40 min, and extracted into the polyethylene glycol phase on the surface of the sample solution.

Following the success of advances in fermentation technology and downstream penicillin production, over the course of the next 40 years, many more β-lactam antibiotics have been discovered and several are presented next.

9.4 CEPHALOSPORIN

In 1948, Giuseppe Brotzu discovered and isolated the first chemical compounds of the cephalosporin group from *Cephalosporium acremonium*. Today, cephalosporin

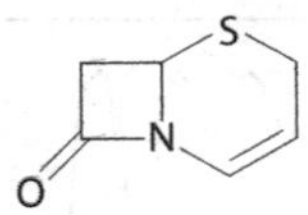

FIGURE 9.5 The backbone structure of cephalosporin is the cephem.

antibiotics are classified into four classes based on their spectrum of activity, resistance to β-lactamase, and their potency differences against gram-positive/negative organisms. The important difference between penicillin and cephalosporin structures can be seen in the enlargement from a five-membered to six-membered ring attached to the common β-lactam core. The backbone of the cephalosporin is cephem (Figure 9.5), which consists of a bicycle system with a four-membered β-lactam ring and a hydrocyclothiazide ring. Compared to the five-member hydrothiazole ring, the cephalosporins exhibit less ring strain than penicillin. Hence, the potency is relatively lower than penicillin. However, they are more stable under acidic conditions and exhibit fewer allergic reactions; these cephalosporins have a prominent place in antibiotic therapy in modern times.

9.5 ISOLATION OF CEPHALOSPORIN C

The structure of cephalosporin C is shown in Figure 9.6. Isolation from fermentation broth is achieved using macroporous, reverse phase, nonionic adsorption resins. One of the commercial resins is known as the Amberlite XAD-n resins (where n varies from 2 to 16 depending on crosslinking). The filtered broth is poured through the XAD resin, which adsorbs the cephalosporins from the solution. The cephalosporin compound is then eluted from the resin with an aqueous solution containing an anionic surface active agent, followed by lyophilization, precipitation, and crystallization. The advantage of using adsorption resins is the high efficiency although it adds cost to large-scale production.

9.6 MONOBACTAMS

In a desire to find ever more useful ß-lactam compounds, the pharmaceutical industry in particular devised new screening techniques and new sources of microbial producers. One outcome was the discovery of the monobactams. The term *monobactam* described the novel group of monocyclic bacterially produced ß-lactam antibiotics having a simple core structure, characterized by the 2-oxoazetidine-1-sulfonic

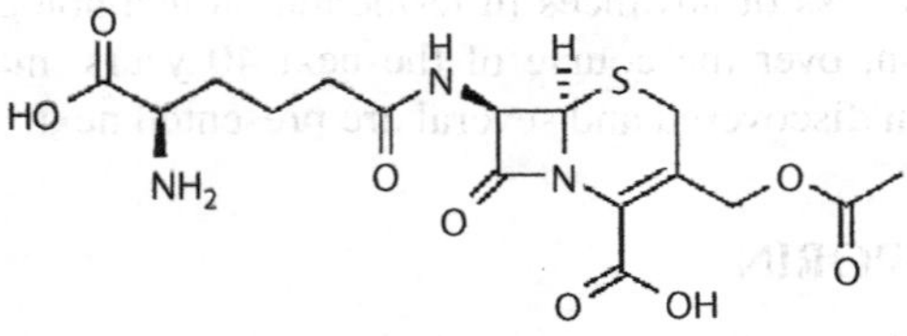

FIGURE 9.6 Structure of cephalosporin C.

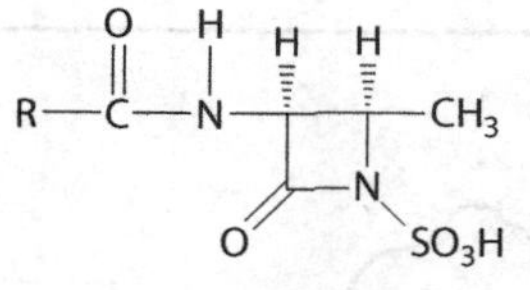

Monobactam

FIGURE 9.7 The core monobactam chemical structure, R represents various amino acid side chains.

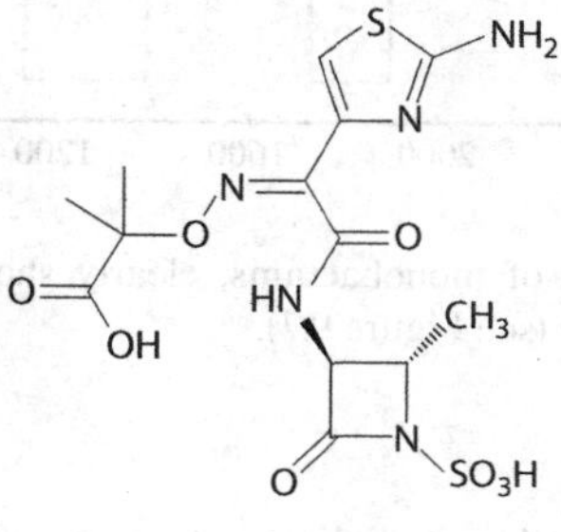

FIGURE 9.8 The commercial monobactam Aztreonam (Azactam).

acid moiety (Figure 9.7). In 1981, researchers in Japan and in the United States independently discovered the monobactams. These compounds were discovered from soil bacteria and detected using supersensitive antimicrobial screens with *Pseudomonas aeruginosa* and *Escherichia coli*.

The first monobactams reported were given the names sulfazecin and iso-sulfazecin (N-acyl derivatives of (S)-3-amino-2-oxo-1-azetidine sulfonic acid (3-aminomonobactamic acid)) produced in fermentation from bacterial strains and not from the more common sources: fungi or actinomycetes. Subsequently, over a 5-year period, 14 naturally occurring monobactams were isolated and characterized from gram-negative bacteria. These β-lactams produced no side effects: however, they only possessed a narrow spectrum of activity toward gram-negative organisms, in contrast to the broad spectrum penicillins and cephalosporins.

These naturally occurring compounds led to the synthesis of many completely synthetic monobactams. For example, Aztreonam is a totally synthetic analog of the naturally occurring monobactams (Figure 9.8).

9.7 STRUCTURAL FEATURES

In the IR spectra, in addition to the diagnostic v_{max} (cm^{-1}) C=O lactam and amide stretches at 1780 cm^{-1} and 1645 cm^{-1}, respectively, there are strong well resolved S=O band stretches around 1240 cm^{-1}, reflecting the presence of sulfonic acid (Figure 9.9). In the mass spectrum a characteristic loss of 80 mu [MH-SO$_3$] from the cleavage of the N-SO$_3$ bond is generally observed, and is a diagnostic fragment ion.

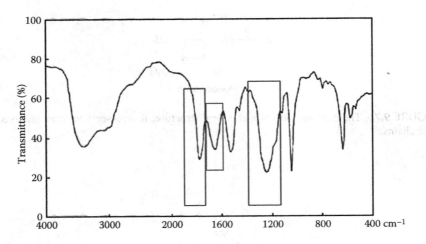

FIGURE 9.9 The IR spectra of monobactams, clearly showing C=O + S=O stretching bands as shown in the structure (see Figure 9.7).

9.8 NOCARDICINS

Related to the monobactams, the nocardicins also contain a core β-lactam but exhibit a ring opened attachment with no sulfur in the molecule. The nocardicin antibiotics were first reported by researchers at Fujisawa Pharmaceutical Co., Japan, in 1976, from a producing strain of *Nocardia uniformis* subsp. *Tsuyamanensis* ATCC 21806. These β-lactams are N-acyl derivatives of 3-amino-nocardicinic acid (Figure 9.10). In total, seven nocardicins were isolated from the metabolites of *Nocardia uniformis*, named nocardicins A-G. Nocardicin A is the major component and also has the highest activity (Figure 9.11).

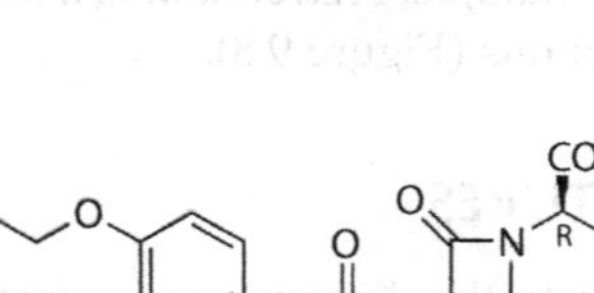

FIGURE 9.10 The structure of 3-aminonocardicinic acid.

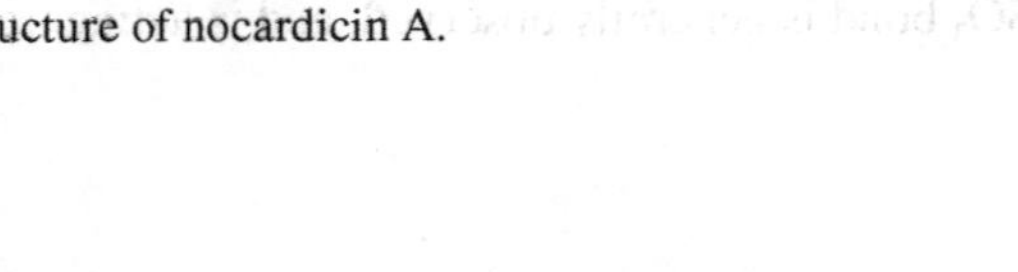

FIGURE 9.11 Structure of nocardicin A.

Both nocardicins and monobactams possess similar mechanisms of action as other β-lactam antibiotics and act as imitators of penicillin-binding proteins, PBPs, in bacteria cell wall formation. Although no valuable antibiotic based on modifying nocardicin A was found, its simple structure without a bicyclic ring core structure indicated that a bicyclic core structure found in the more traditional β-lactam antibiotics may not be necessary. The monocyclic nucleus only has one four-membered ring, which has less ring strain in comparison to penicillin. A less rigid structure leads to the lower activity of the β-lactam ring. As a result, norcardicin A possesses only moderate activity *in vitro* against some gram-negative bacteria. It is a narrow spectrum antibiotic.

9.9   ISOLATION

Nocardicin A can be obtained directly from the crude fermentation broth as an antibiotic mixture using the following scheme and then recrystallized from aqueous ethanol. In this example, a macroporous resin, Diaion HP20 was used in a similar way to XAD resins, which absorb the crude antibiotic mixtures onto the resin matrix.

Nocardia uniformis Broth Filtrate, pH 4

  Sorb on Diaion HP20 resin and elute with MeOH-H2O, 3:7.

  Concentrate *in vacuo* and acidify to pH 2.5 (HCl).

Crude antibiotic mixture

  Chromatograph on Diaion HP20 resin

  Adj to pH 7 (NaOH)

  1. Elute with 3% NaCl

Nocardicin A

  2. Continue elute with MeOH-H2O, 3:7, followed by concentration and acidification to pH 2.3 (HCl)

Nocardicin B

9.10   CARBAPENEM

The simplest naturally occurring carbapenem antibiotic has the core structure shown in Figure 9.12. Other carbapenem antibiotics are based on this core, with substituted groups on C2 and C6. The first carbapenem antibiotic, thienamycin (Figure 9.13), was also first discovered in 1976, isolated from the fermentation broth of *Streptomyces*

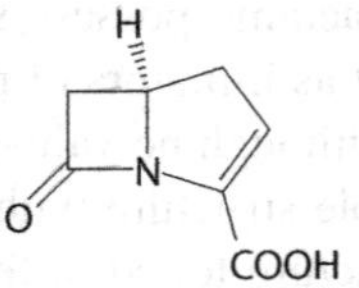

FIGURE 9.12   The core carbapenem structure.

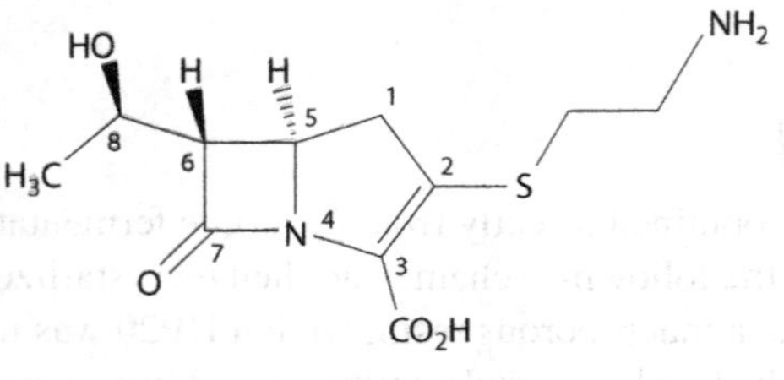

FIGURE 9.13   The structure of thienamycin.

cattleya. Thienamycin has high potency, broad spectrum, antibacterial activity, and relatively high resistance to β-lactamases. The discovery of thienamycin represented a new family of β-lactam antibiotics and more than 40 natural carbapenem antibiotics have been isolated.

Most of the naturally occurring carbapenem antibiotics have a 1-hydroxyethyl group on C6. The configuration differences at the chiral center, C8, result in epimers. Due to the low isolation efficiency and multiple products formed under fermentation conditions, all carbapenem antibiotics for clinic use are produced by total synthesis, although this leads to a high cost compared to producing penicillin and its derivatives directly from fermentation.

Although thienamycin is high in potency and resistant to β-lactamases, its instability is a limiting factor in clinic use. Thus, more chemically modified carbapenem antibiotics were synthesized and several carbapenem antibiotics, for example, imipenem (Figure 9.14), have been marketed.

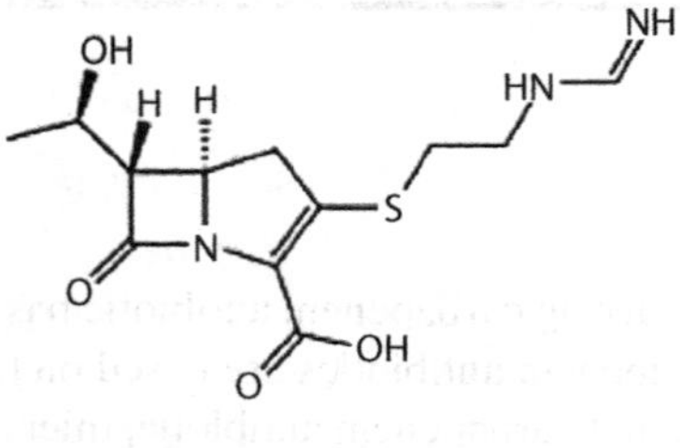

FIGURE 9.14   The structure of imipenem.

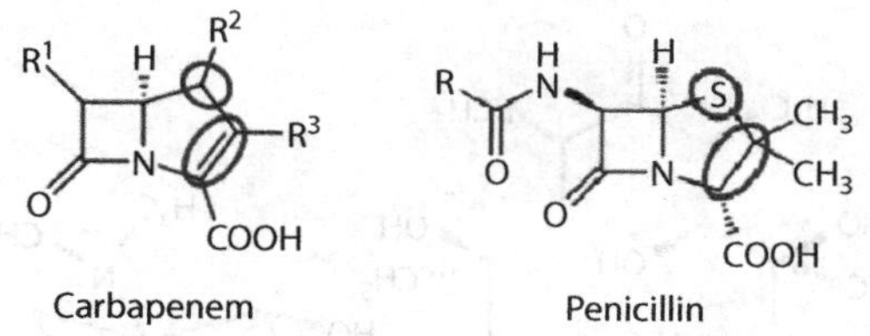

FIGURE 9.15 Structural features of the rings comparing carbapenem to penicillins.

9.11 STRUCTURE DETERMINATION AND STABILITY OF CARBAPENEM

The reasons for the higher reactivity of the β-lactam ring in carbapenems is explained by differences in their structure shown in Figure 9.15. The carbon atom substituted sulfur bonding in penicillin results in a more rigid five-membered ring. When a double bond is formed between C2 and C3 and conjugated with the lone pairs on the nitrogen atom, this conjugation forces the original "envelope-like" bicyclic system to be on the same plane. This leads to further ring strain, which in turn increases the β-lactam activity. However, as noted, a problem of stability occurs. For final structure determination a stable derivative of thienamycin was prepared of the N-acetyl methyl ester by crystallization from acetonitrile–benzene–hexane and suitable for X-ray crystallographic analysis.

9.12 ISOLATION OF THIENAMYCIN

The most difficult part of isolating thienamycin from fermentation broth is the hydrolytic instability of the carbapenem nucleus. The pKa of the carboxylate is around 3.1, while the amino group is larger than 8. It is zwitterionic at neutral pH. Under acidic conditions (pH 2 to 5) instability is similar to penicillin G. As the pH increases, the hydrolysis rate of the lactam bond immediately accelerates.

In the isolation of thienamycin, UV monitoring coupled to high-performance liquid chromatography (HPLC) was effective in detecting the presence of three carbapenems with the most abundant being thienamycin (UV λ_{max} 297nm).

The researchers took advantage of the zwitterionic nature of this carbapenem and applied ion exchange column chromatography. The application of a weak acid anion exchange column Dowex 1x2 (HCO_3^-) with elution using cold carbonic acid as the eluent at a pH ~5 was successful.

Another approach, since the target compound is zwitterionic, is to use an ionic extraction for the first step from the fermentation broth. The advantage of ionic extraction with an ion-pairing reagent is the feasibility of large-scale processing using very short residence times for thienamycin at unfavorable pH values. Furthermore, ion exchange chromatography is not practical for large-scale isolation due to the cost. The ion-pairing extraction provides a cheaper and even more effective way in the isolation of thienamycin. The ion-pairing reagent chosen was dinonylnaphthalene sulfonate (DNNS) from acidic broth and methyl tricapryl ammonium bicarbonate from basic broth.

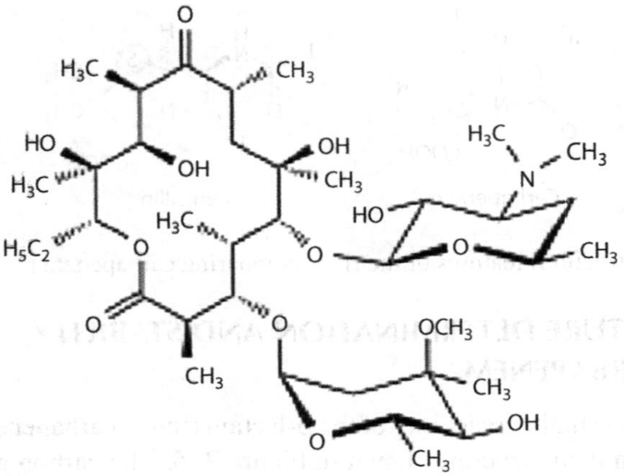

FIGURE 9.16 Structure of erthromycin.

9.13 ANTIBIOTIC MACROLIDES: ERYTHROMYCIN

Erythromycin (Figure 9.16) is a macrolide antibiotic that has an antimicrobial spectrum similar to or slightly wider than that of penicillin, and is often prescribed for people who have an allergy to penicillins. For respiratory tract infections, it has better coverage of atypical organisms, including *Mycoplasma* and legionellosis.

This macrocyclic compound contains a 14-membered lactone ring with ten asymmetric centers and two sugars (L-cladinose and D-desosamine), making it a compound that is difficult to produce via synthetic methods. Erythromycin is produced in fermentation from a strain of the actinomycete *Saccharopolyspora erythraea*.

9.14 ANTIPARASITIC DRUGS: AVERMECTINS

The avermectins (Figure 9.17) form a series of 16-membered macrocyclic lactone derivatives with potent anthelmintic and insecticidal properties. These naturally

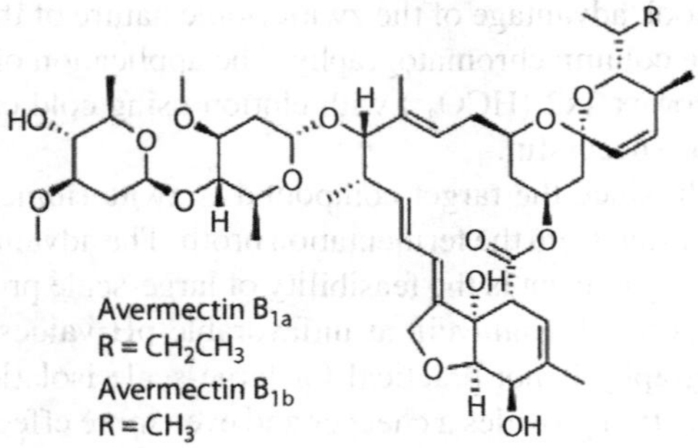

Avermectin B_{1a}
$R = CH_2CH_3$
Avermectin B_{1b}
$R = CH_3$

FIGURE 9.17 The avermectin family of compounds.

HISTORICAL NOTE

The avermectin story is a research collaboration between scientists in two countries. In 1978, an actinomycete was isolated at The Kitasato Institute, Tokyo, Japan from a soil sample collected at Kawana, Japan, and the isolated actinomycete was sent to Merck Sharp and Dohme for testing. Early tests indicated activity against *Nematospiroides dubis* in mice without toxicity. Subsequently, the anthelmintic activity was isolated and identified as a family of closely related compounds in 1978.

occurring compounds are generated as fermentation products by *Streptomyces avermitilis*, a soil actinomycete. Eight different avermectins were isolated as four pairs of homologous compounds, with a major (a-component) and minor (b-component) component, usually in ratios of 80:20 to 90:10.

9.15 TETRACYCLINES

Tetracycline compounds form another group of broad spectrum antibiotics whose general usefulness has been reduced with the onset of bacterial resistance. They are defined as a subclass of polyketides having an octahydrotetracene-2-carboxamide skeleton (Figure 9.18). Tetracyclines are generally used in the treatment of infections of the urinary tract and the intestines, and are used in the treatment of infections caused by chlamydia, especially in patients allergic to β-lactams and macrolides. However, their use for these indications has decreased due to widespread development of drug resistance to these compounds. Their most common current use is in the treatment of moderately severe acne and rosacea. In addition, they may be used to treat Legionnaires' disease. They also are used in veterinary medicine, particularly on swine.

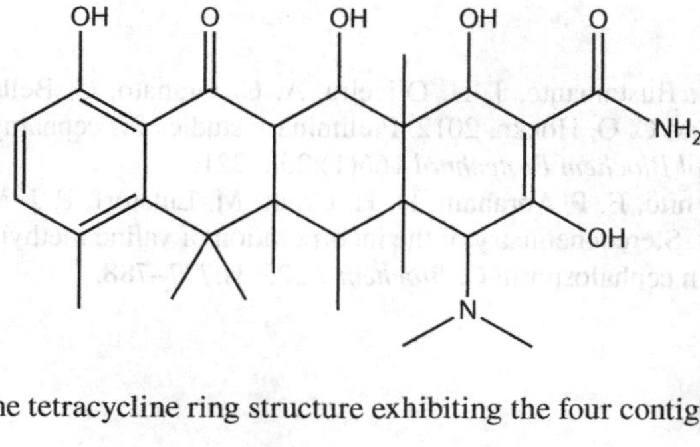

FIGURE 9.18   The tetracycline ring structure exhibiting the four contiguous rings.

SUMMARY

In this chapter, we give examples of several important naturally produced anti-infective compounds. These are compounds are produced by microorganisms under fermented conditions. They include the β-lactam antibiotics (e.g., penicillin, cephalosporin), erythromycin, the avermectins, and tetracyclines.

QUESTIONS

1. Why is IR spectroscopy useful for penicillin compounds?
2. Why is IR spectroscopy more useful than UV for penicillin compounds?
3. The monobactams contain an SO3 function. How would you establish it has an N-SO3 and not an O-SO3 or C-SO3 bond?
4. What are the ways to identify the sulfur atom in a molecule?

FURTHER READING

A. G. Brown. 1987. Discovery and development of new β-lactam antibiotics. *Pure Appl Chem* 59(3):475–484.

K. Schugerl. 2005. Extraction of primary and secondary metabolites. *Adv Biochem Eng Biotechnol* 92:1–48.

G. H. Wagman. 1989. *Natural Products Isolation: Separation Methods for Antimicrobials, Antivirals and Enzyme Inhibitors.* Elsevier Science Limited.

PENICILLINS

P. Y. Bi, H. R. Dong, and Q. Z. Guo. 2009. Separation and purification of penicillin G from fermentation broth by solvent sublation. *Separation and Purification Technology* 65(2):228–231.

P.-Y. Bi, D.-Q. Li, and H.-R. Dong. 2009. A novel technique for the separation and concentration of penicillin G from fermentation broth: Aqueous two-phase flotation. *Separation and Purification Technology* 69(2):205–209.

Q. Liu, J. Yu, W. Li, X. Hu, H. Xia, H. Liu, and P. Yang. 2006. Partitioning behavior of penicillin G in aqueous two phase system formed by ionic liquids and phosphate. *Separation Sci Tech* 41(12):2849–2858.

Y. Lu and X. Zhu. 2001. Solvent sublation: Theory and application. *Separation Purification Rev* 30(2):157–189.

CEPHALOSPORINS

B. Neto Ade, M. C. Bustamante, J. H. Oliveira, A. C. Granato, C. Bellao, A. C. Badino, M. Barboza, and C. O. Hokka. 2012. Preliminary studies for cephamycin C purification technique. *Appl Biochem Biotechnol* 166(1):208–221.

C. P. Pang, R. L. White, E. P. Abraham, D. H. Crout, M. Lutstorf, P. J. Morgan, and A. E. Derome. 1984. Stereochemistry of the incorporation of valine methyl groups into methylene groups in cephalosporin C. *Biochem J* 222(3):777–788.

CARBAPENEMS

G. Albers-Schoenberg, B. H. Arison, O. D. Hensens, J. Hirshfield, K. Hoogsteen, E. A. Kaczka, R. E. Rhodes, J. S. Kahan, and F. M. Kahan. 1978. Structure and absolute configuration of thienamycin. *J Am Chem Soc* 100(20):6491–6499.

J. S. Kahan, F. M. Kahan, R. Goegelman, S. A. Currie, M. Jackson, E. O. Stapley, T. W. Miller, et al. 1979. Thienamycin, a new beta-lactam antibiotic. I. Discovery, taxonomy, isolation and physical properties. *J Antibiot (Tokyo)* 32(1):1–12.

K. M. Papp-Wallace, A. Endimiani, M. A. Taracila, and R. A. Bonomo. 2011. Carbapenems: Past, present, and future. *Antimicrob Agents Chemother* 55(11):4943–4960.

S. S. Weaver, G. P. Bodey, and B. M. LeBlanc. 1979. Thienamycin: New beta-lactam antibiotic with potent broad-spectrum activity. *Antimicrob Agents Chemother* 15 (4):518–521.

MONOBACTAMS

R. Cooper. 1983. Chemical spectroscopic characterization of monobactams. *J Antibiotics* 36:1258–1262.

R. Cooper, K. Bush, P. A. Principe, W. H. Trejo, J. S. Wells, and R. B. Sykes. 1983. Two new monobactam antibiotics produced by a *Flexibacter* sp. *J Antibiotics* 36:1252–1257.

10 Terpenes in Human Health

10.1 THE STATIN DRUGS

In this chapter we return to some more examples of terpenes, particularly those offering human health benefits. One important group of terpenes forms the statin family of cholesterol-lowering drugs. Statins inhibit the enzyme hydroxymethyl-glutaryl-CoA (HMG-CoA) reductase inhibitors known as monacolins, which play an important role in the liver to produce cholesterol. Over the years, a great deal of research has resulted in a number of statin compounds reaching the market, including Mevacor, chemically known as lovastatin or monacolin K (Figure 10.1) produced by Merck from a microbial source, and later Simvastatin originating at Bristol Myers Squibb. The best-selling statin drug to date is sold by Pfizer and known as Lipitor.

Another source of statins is found in red yeast rice (RYR), a traditional Chinese food and medicine, produced from fermented rice using a strain of the food fungus *Monascus purpureus* Went Rice. It is sold in China as a drug and in the West as a dietary supplement. This product has been widely used as a food supplement in China and other Asian countries for centuries for the preparation of fish, meat, and bean curd, and in the making of rice wine. It was used in traditional Chinese food and medicine "to improve blood circulation and to promote a healthy digestive function in the body." These same compounds are also found in the oyster mushroom (Figure 10.2). In RYR, lovastatin is most often produced as the major component, but several other related monacolin analogs are generally coproduced in fermentation. The structures of these statin (monacolin)-related drugs are shown in Figure 10.3.

Modern methods of analysis are used to separate and identify the monacolins. Using a reverse phase C18 (5 μm) column as a stationary phase, separation is achieved with a linear gradient of aqueous acetonitrile and 0.1% trifluoroacetic acid (TFA) as the mobile phase at a flow rate of 0.9 mL/min. Detection of compounds is achieved by coupling the system either to a photodiode array (PDA) detector or, for liquid chromatography–mass spectrometry (LC-MS) analysis, to an electrospray ionization (ESI) quadrupole ion-trap detector.

10.2 STEROIDS AND CORTISONE

A very important and large class of naturally occurring compounds is the steroid, represented in Figure 10.4. One example is cortisone, a 17-hydroxy-11-dehydrocorticosterone steroid hormone (Figure 10.5). It is used to treat a variety of ailments, particularly to reduce inflammation, eczema, and dermatitis.

Isolation of steroids may be achieved with the use of organic solvents. Often a saponification step is required using alcohol to rupture the steroid ester bond, releasing the steroid, which can be extracted into either hexane or ether.

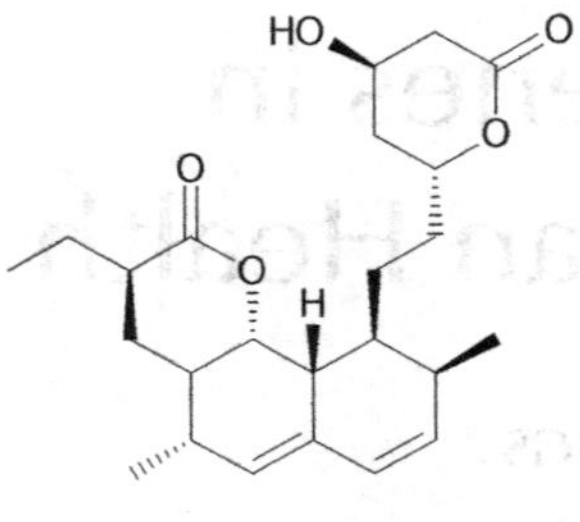

FIGURE 10.1   The structure of the important statin monacolin K.

FIGURE 10.2   The oyster mushroom, *Pleurotus ostreatus*, growing on the tree bark produces monacolin-like compounds. (Courtesy of Steven Foster.)

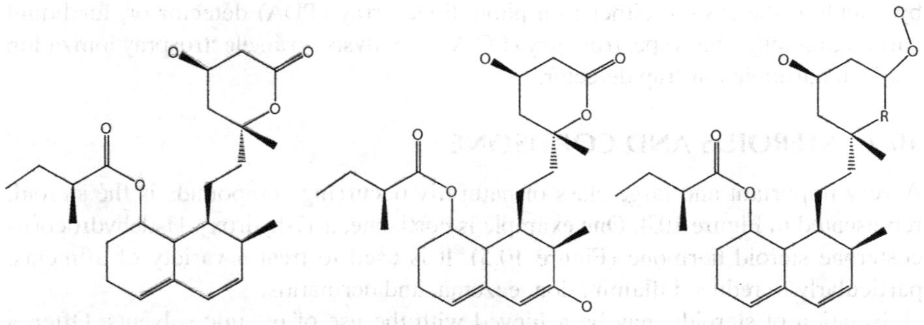

Compactin

6-Hydroxy-iso-compactin

5'-Phosphocompactin acid
R=Phosphate moiety

FIGURE 10.3   Structures of statin (monacolin) homologs isolated from natural sources.

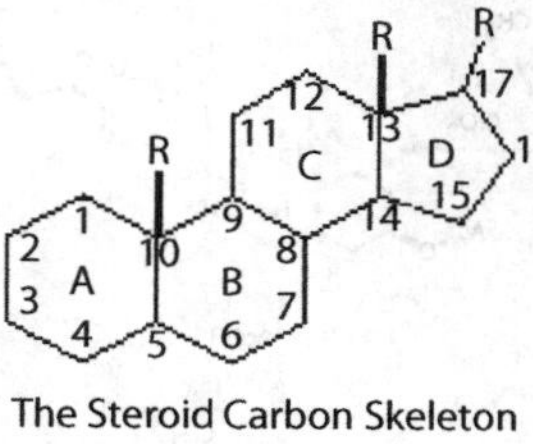

FIGURE 10.4   The steroid skeleton.

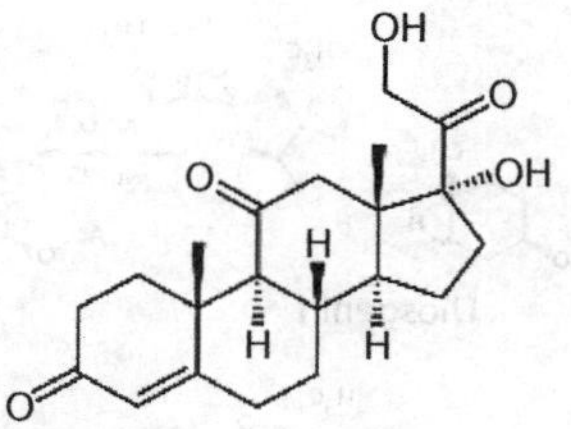

FIGURE 10.5   The structure of a medically useful steroid cortisone.

10.3   STEROIDS FROM YAMS

The genesis of the birth control pill can be traced back to wild yams growing in Mexico. The wild yam is a member of the Dioscoreaceae family comprising hundreds of species. However, only four species are of relevance for medicinal purposes: *Dioscorea villosa* (found in the United States); *Dioscorea opposita* and *Dioscorea hypoglauca* (native to Asia); and the Mexican yam, *Dioscorea barbasco.*

The oral contraceptive pill is a combination of an estrogen (estradiol) and a progestogen (progestin), and when females ingest these pills they inhibit fertility. They were first approved for contraceptive use in the United States in 1960. Currently, more than 100 million women use them worldwide.

The primary active chemical agent in wild yam is the steroidal saponin, known as diosgenin. It is actually present in the roots of the wild yam as dioscin, a steroidal saponin whose aglycone is diosgenin. The glycoside dioscin from the Mexican wild yam root, *Dioscorea*, constituted the first significant sapogenin plant source for steroid drugs and this compound can be chemically converted to the hormone progesterone (Figure 10.6). The chemical steps of conversion from diosgenin to progesterone are shown in Figure 10.7.

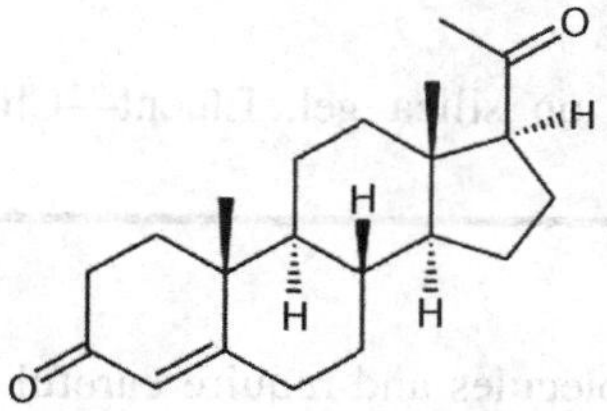

FIGURE 10.6   The steroidal hormone progesterone.

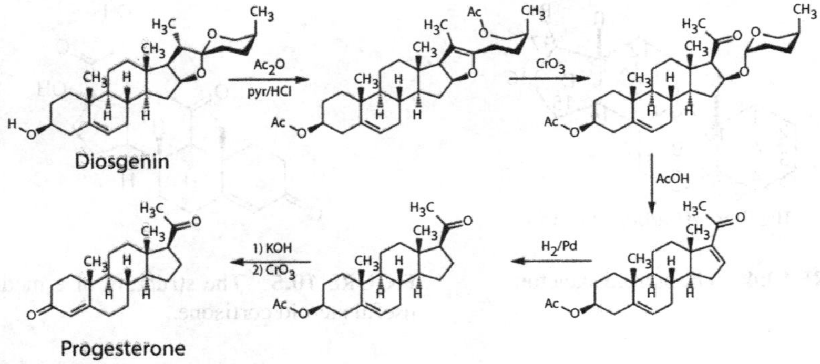

FIGURE 10.7 The synthetics steps of the conversion of the steroid aglycone from the yam: diosgenin into progesterone.

Isolation of Diosgenin Leading to Progesterone

Pretreatment

 (Remove soil and microbes; pulverized)

Primary extraction

 Water (hydrophilic) + Ether

2nd extraction

 n-Butanol (dioscin) + Water

Column chromatography

 Purified dioscin

Hydrolysis

⇩ Diosgenin

Marker degradation

⇩ Progesterone

Column chromatography on silica gel, Eluent—Chloroform:Methanol:Water (7:1.5:0.1) mixture

---

Steroids are complex molecules and require careful spectroscopic analysis. For example, the ^{13}C NMR of diosgenin, shown in Figure 10.8, reveals all the carbon resonance signals. The infrared (IR) spectrum is shown in Figure 10.9.

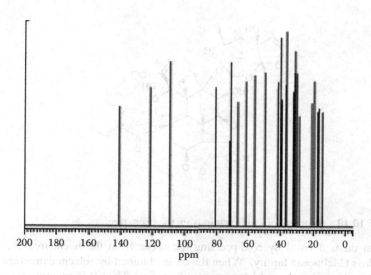

FIGURE 10.8   Example of a ^{13}C NMR spectrum of the steroid, diosgenin.

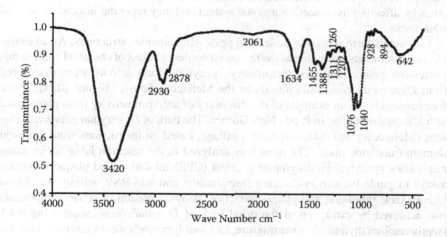

FIGURE 10.9   Typical IR spectrum of the steroid, diosgenin.

10.4   NEEM OIL AND OTHER LIMONOIDS

The neem tree is a fast growing tree native to South Asia, and is grown on a large scale in India. The tree has been used for medicinal purposes and pest control in India for thousands of years. It is used in Ayurvedic medicine to reduce fevers and has been shown to contain effective antimalarial (*Plasmodium falciparum*) compounds. The bark and leaves have several uses: antiseptic, antiviral, anti-inflammatory, antiulcer, and antifungal. Neem oil is also effective as a repellant (insecticide), miticide, and fungicide, and functions as an antifeedant, which discourages insects' feeding patterns.

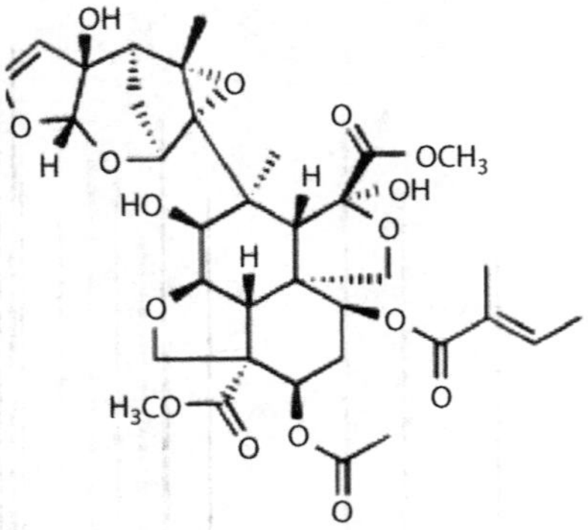

FIGURE 10.10 Structure of azadirachtin from the neem tree.

Neem oil is derived by cold-pressing the seeds from the neem tree, *Azadiracta indica* Juss (Meliaceae family). When the oil is obtained by solvent extraction it gives a lower quality oil and used in soap manufacturing. The oil is mainly comprised of triglycerides and large amounts of triterpenoid compounds (limonoids), of which azadirachtin (Figure 10.10) is well studied. The limonoids appear to have insecticidal properties by affecting the insect's hormonal system and they repel the insects' larvae and adult forms.

Limonoids form a class of molecules possessing complex structures. As an example, new limonoids belonging to the tetranortriterpenoid class of chemical compounds possessing potent anti-RSV (respiratory syncytial virus) activity were discovered from *Dysoxylum gaudichaudianum* in the Meliaceae family (Figure 10.11). These four limonoids are an example of the discovery of antiviral activity from plants used as traditional medicine in Papua New Guinea. The bark of *Dysoxylum gaudichaudianum* (Meliaceae) was collected from a village, based on instructions from the local shaman (medicine man). The plant was analyzed in the research laboratories using respiratory syncytial viral cytopathic effect (CPE) inhibition and plaque reduction assays to guide bioactivity-directed fractionation and anti-RSV activity was found in both CPE inhibition and plaque reduction assays. Isolation of active components was achieved by extraction of the dried bark of *D. gaudichaudianum* using a 1:1 isopropanol/methylene chloride mixture, and four limonoids were eventually isolated in pure form.

As an example of a structure determination, this was achieved based on a combination of spectroscopic techniques and comparison to a known terpene skeleton, namely trichilinin E. Limonoid A has a molecular formula of $C_{35}H_{48}O_{10}$ that can be determined from high-resolution mass spectrometry (HRMS) to allow an accurate molecular formula determination. The values were obtained from electrospray HRMS: M+, m/z 628.3234, and the calculated value is 628.3248, in agreement with the proposed molecular formula. In the ^{13}C NMR, the following 35 carbon resonance signals were observed and assigned as follows: 8 methyl (CH3), 6 methylene (CH2), 11 methine (CH) carbons, and therefore 10 quaternary (C) carbons in the molecule and consistent with the proposed structure.

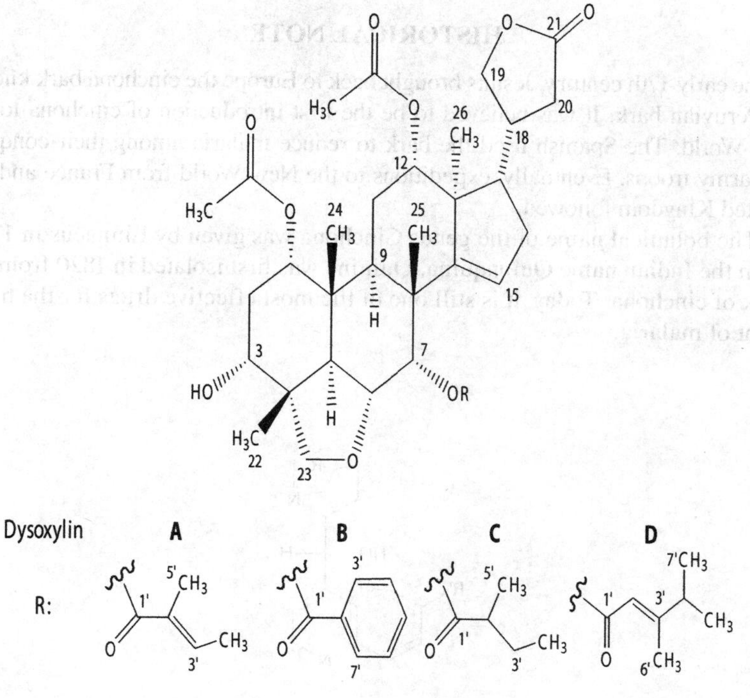

Dysoxylin

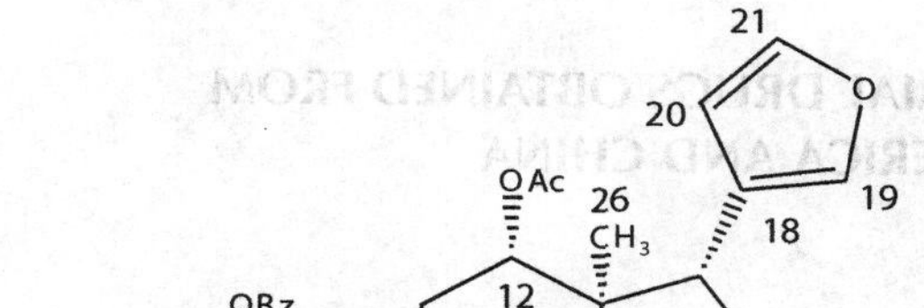

FIGURE 10.11 Example of naturally occurring limonoids from plants in Papau New Guinea.

HISTORICAL NOTE

In the early 17th century, Jesuits brought back to Europe the cinchona bark known as Peruvian bark. It was believed to be the first introduction of cinchona to the Old World. The Spanish used the bark to reduce malaria among their conquering army troops. Eventually, expeditions to the New World from France and the United Kingdom followed.

The botanical name of the genus Cinchona was given by Linnaeus in 1742, from the Indian name Quinaquina. Quinine was first isolated in 1820 from the bark of cinchona. Today, it is still one of the most effective drugs for the treatment of malaria.

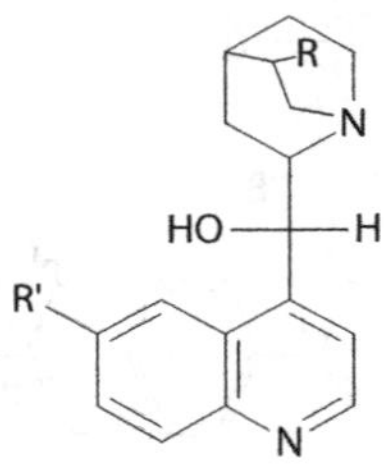

FIGURE 10.12   The structure of quinine, the antimalarial drug from the cinchona bark.

10.5   ANTIMALARIAL DRUGS OBTAINED FROM SOUTH AMERICA AND CHINA

10.5.1   QUININE

Another example of a drug derived from a plant source is quinine from the South American cinchona bark, *Cinchona officinalis*, to treat malarial diseases. Quinine (Figure 10.12) is an antifever agent and is especially important in treating malaria. The bioactive compounds include cinchona alkaloids, one of which is quinine, R = vinyl; R′ = methoxyl.

10.5.2   ARTEMISININ

Artemisinin (Figure 10.13) is also an antimalarial drug, discovered in the leaves of *Artemisia annua* (annual wormwood).

10.6   ANTICANCER DRUG: TAXOL FROM THE PACIFIC YEW

Paclitaxel, also known as Taxol® (Figure 10.14), is a diterpenoid. It was first isolated from the bark of Pacific yew trees (*Taxus brevifolia*).

FIGURE 10.13 Artemisinin, a powerful antimalarial drug obtained from the leaves of *Artemisia annua*. (Photograph courtesy of Steven Foster.)

HISTORICAL NOTE

An example of a Traditional Chinese Medicine, the plant *Artemisia annua* has been used by Chinese herbalists for over 2000 years and was believed to be used in the treatment of skin diseases and malaria. One of the earliest records dates to 200 BC, in the "Fifty-Two Prescriptions" and specifically its antimalarial application was first described in the 4th century in *The Handbook of Prescriptions for Emergencies*.

In the 1960s, a research program was set up by the Chinese army to find an adequate treatment for malaria. By 1972, artemisinin was discovered in the leaves of *Artemisia annua*. During the screening of more than 5000 Traditional Chinese Medicines, artemisinin was found most effective and cleared malaria parasites in patients better than all previous drugs. Due to the secret nature of this program, however, the research work was never given the full recognition it deserved.

Paclitaxel   10-DAB

FIGURE 10.14   The structure of paclitaxel from *Taxus brevifolia* and the precursor DAB from *Taxus baccata*. (Photograph courtesy of Steven Foster.)

Paclitaxel is a mitotic inhibitor used in cancer chemotherapy. It was approved by the U.S. Food and Drug Administration (FDA) for treatment of drug-resistant ovarian and breast cancers and also is used in the treatment of Kaposi's sarcoma and lung cancer. Subsequently, it has become a major research tool of study in cancer therapy. Paclitaxel works by stabilizing microtubules, which play important roles in cell division. This stabilization of microtubules inhibits mitosis: the cells are unable to multiply and thus tumors are unable to grow.

HISTORICAL NOTE

The compound taxol was discovered by Monroe and Wall seeking to find anticancer agents and in 1971 they published their findings. Remarkably, it took many years before further study at the National Cancer Institute (NCI) advanced the compound into the clinic. The NCI showed great reluctance to pursue taxol for a number of reasons: the isolation and extraction was difficult, and the yew tree produced only small amounts of compound. Since the bark is a finite resource, once the tree is stripped of its bark it dies, and because of the supply problem taxol did not progress for many years.

However, due to the efforts of Dr. Matthew Suffness at NCI, a fresh review found taxol to be very active against B16 melanoma, a newly adopted tumor model, and thus the program gained momentum. In 1978, taxol showed the ability to cause considerable regression in a mammary tumor xenograft and interest was increased when the mechanism of action in mitotic tubules was discovered.

By the early 1980s, the NCI put out a bid for industry support and Bristol Myers Squibb decided to develop taxol into a drug. New research showed taxol could be obtained from the needles of the yew tree, an ecologically much better prospect. Subsequently, it was discovered as a fungal metabolite with the potential for large-scale production from fermentation.

Although a total synthesis of paclitaxel was independently completed by two groups, it requires almost 40 steps, resulting in a low overall yield and is not economically feasible.

The synthesis of paclitaxel uses an elegant combination of isolating large amounts of a precursor compound related to taxol followed by an additional semisynthetic step to produce the final product. The first source is the English yew *Taxus baccata,* which is available in nature in large quantities where the compound 10-DAB can be obtained from the needles. The difference between 10-DAB and paclitaxel is that 10-DAB has no ester side-chain at C-13. Thus, to complete the taxol structure, the prepared side-chain is then attached to the C-13 hydroxyl group of 10-DAB to obtain paclitaxel on a large scale. In this manner, paclitaxel was manufactured by a semi-synthetic production from a natural precursor.

10.7 BRYOSTATINS

Bryostatin compounds (Figure 10.15) belong to a group of macrolide lactones. They were found in a marine species of bryozoan, *Bugula neritina*. The structure of bryostatin 1 was determined in 1982 and to date 20 different bryostatins have been isolated. They are potent modulators of protein kinase C and are currently under investigation as anticancer agents.

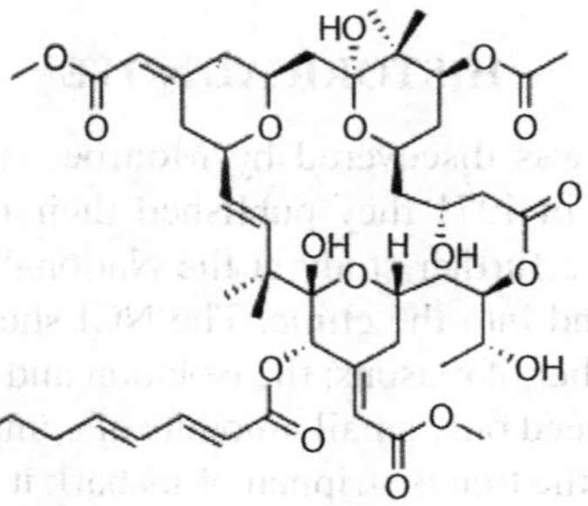

FIGURE 10.15   Structure of a bryostatin from marine sources.

SUMMARY

This chapter provides various important natural product contributors to human health: the terpene-related monacolin compounds and the steroids, including cortisone and progesterone. Various natural product examples from terrestrial, microbial, and marine sources are provided: the antimalarial quinine and artemisinin; and the anticancer compounds paclitaxel and bryostatin.

QUESTIONS

1. Artemisinin is a complex oxygenated terpene. Describe the types of oxygen bonds in the molecule.
2. What is the best approach to determine the dioxygen bridge structure in the artemisinin molecule?

FURTHER READING

MONACOLINS

A. W. Alberts, J. Chen, G. Kuron, V. Hunt, J. Huff et al. 1980. Mevinolin: A highly potent competitive inhibitor of hydroxymethlglutaryl-coenzyme A reductase and a cholesterol-lowering agent. *Proc Natl Acad Sci USA* 77(7):3957–3961

P. Bobek, O. Ozdín, and M. Mikus. 1995. Dietary oyster mushroom (*Pleurotus ostreatus*) accelerates plasma cholesterol turnover in hypercholesterolaemic rat. *Physiol Res* 44(5):287–291.

A. Endo, M. Kuroda, and Y. Tsujita. 1976. ML-236A, ML-236B, and ML-236C, new inhibitors of cholesterogenesis produced by *Penicillium citrinium*. *J of Antibiotics* 29(12):1346–1348.

Y. L. Lin, T. H. Wang, M. H. Lee, and N. W. Su. 2008. Biologically active components and nutraceuticals in the Monascus-fermented rice: A review. *Applied Microbiology and Biotechnology* 77(5):965–973.

Steroids

American Chemical Society and Sociedad Química de México. 1999. *The "Marker Degradation" and Creation of the Mexican Steroid Hormone Industry, 1938–1945: An International Historic Chemical Landmark.* American Chemical Society.

C. Djerassi. 2011. The pill at 50 (in Germany): Thriving or surviving? *Journal für Reproduktionsmedizin und Endokrinologie* 8 (1):14–31.

S. P. Mishra and V. G. Gaikar. 2004. Recovery of diosgenin from dioscorea rhizomes using aqueous hydrotropic solutions of sodium cumene sulfonate. *Ind Eng Chem Res* 43(17):5339–5346.

Limonoids

J. Butterworth and E. Morgan. 1968. Isolation of a substance that suppresses feeding in locusts. *Chemical Communications* 1:23.

J. L. Chen, M. R. Kernan, S. D. Jolad, C. A. Stoddart, M. Bogan, and R. Cooper. 2007. Dysoxylins A-D, tetranortriterpenoids with potent anti-RSV activity from *Dysoxylum gaudichaudianum. J Natural Products* 71:312–315.

I. L. Musza, L. M. Killar, P. Speight, C. J. Barrow, A. M. Gillum, and R. Cooper. 1995. Minor liminoids from *Trichilia rubra. Phytochemistry*, 39:621–624.

M. Nakatani. 1999. *Azadirachtins* in *The Biology-Chemistry Interface: A Tribute to Koji Nakanishi*, R. Cooper and J. Snyder (Eds.). Marcel Dekker.

Quinine and Artemisinin

M. Willcox, G. Bodeker, P. Rasoanaivo, and J. Addae-Kyereme. 2004. *Traditional Medicinal Plants and Malaria.* Taylor & Francis.

Taxol

R. A. Holton, C. Somoza, H. B. Kim, F. Liang, R. J. Biediger, P. D. Boatman, M. Shindo, C. C. Smith, and S. Kim. 1994. First total synthesis of taxol. 1. Functionalization of the B ring. *J Am Chem Soc* 116(4):1597–1598.

K. C. Nicolaou, Z. Yang, J. J. Liu, H. Ueno, P. G. Nantermet, R. K. Guy, C. F. Claiborne, et al. 1994. Total synthesis of taxol. *Nature* 367(6464):630–634.

A. Stierle, G. Strobel, and D. Stierle. 1993. Taxol and taxane production by Taxomyces andreanae, an endophytic fungus of Pacific yew. *Science* 260(5105):214–216.

M. C. Wani, H. L. Taylor, M. E. Wall, P. Coggon, and A. T. McPhail. 1971. Plant antitumor agents. VI. The isolation and structure of taxol, a novel antileukemic and antitumor agent from *Taxus brevifolia. J Am Chem Soc* 93 (9):2325–2327.

Bryostatins

K. J. Hale and S. Manaviazar. 2010. New approaches to the total synthesis of the bryostatin antitumor macrolides. *Chem Asian J* 5(4):704–754.

G. R. Pettit, Y. Kamano, and C. L. Herald. 1986. Antineoplastic agents, 118. Isolation and structure of bryostatin 9. *J Nat Prod* 49(4):661–664.

11 Carotenoids

Carotenoids are also terpenoids, belonging to a very large family of organic pigmented compounds. They are tetraterpenoids: terpenoids of eight isoprene units (discussed in Chapter 4), possessing 40 carbons within the molecular skeleton, as shown in Figure 11.1.

Carotenoids are of importance to both the plant and the host microorganism. Carotenoids are typically found in photosynthetic plants or fungi, algae, and animal products. They are also found in eggs, animal tissues, fruits, and many vegetables. Carotenoids are involved in the photosynthetic process and offer protection against photodamage. They provide the yellow, orange, and red colors to their respective fruits and vegetables. Carotenoids are one of the two key pigments that contribute to the skin yellowness of humans. They are also recognized as valuable nutritional compounds to the human body. They are essential for enhancing the immune system, possess photoprotection ability, and may support reproductive health.

β-Carotene in the body is a form of provitamin A and acts as an antioxidant considered beneficial in preventing human diseases. The carotene is cleaved by the enzyme β-carotene-15,15′-dioxygenase to form retinol (Figure 11.2).

The more than 600 known carotenoids can be divided into two categories: xanthophylls and carotenes. Xanthophylls are those derivatives containing oxygen atoms, whereas carotenes only contain carbon and hydrogen. In this chapter, four carotenoids are described in detail: β-carotene (provitamin A), lutein, zeaxanthin (for eye health), and lycopene (for cardio health).

HISTORICAL NOTE

The earliest studies on carotenoids date to the early 19th century. Although at the time many carotenoids were discovered and named, their structures were not known. As an example, in 1831, carotene was discovered and isolated by the scientist Wackenroder as a by-product of the search for a medical agent— an anthelminthic (ridding the body of parasitic worms). In 1837, Berzelius named the yellow pigments obtained from autumn leaves as xanthophylls. By the 1860s, lutein was isolated, followed 10 years later by the identification of carotene in animal tissues. The first separation and purification of carotenes and xanthophylls via chromatography was achieved around 1910 by Willstatter and Mieg, who first established the empirical formula of $C_{40}H_{56}$ for β-carotene, and $C_{40}H_{56}O_2$ for xanthophylls. However, it required 20 more years before Karrer elucidated the chemical structures of carotenoids and demonstrated the transformation of carotenoids into vitamin A in the body.

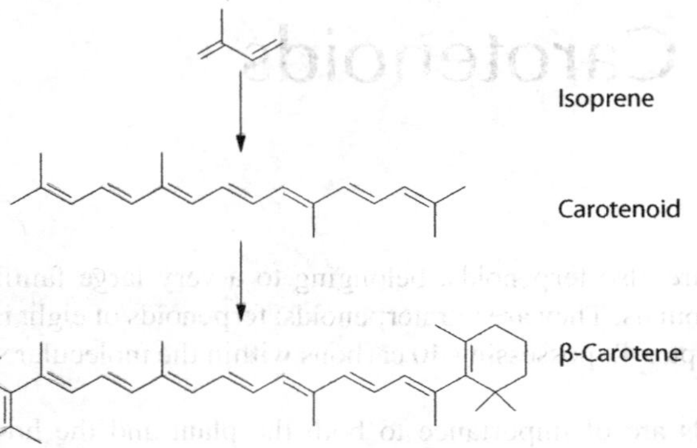

FIGURE 11.1   Beginning from an isoprene unit, the pathway to β-carotene.

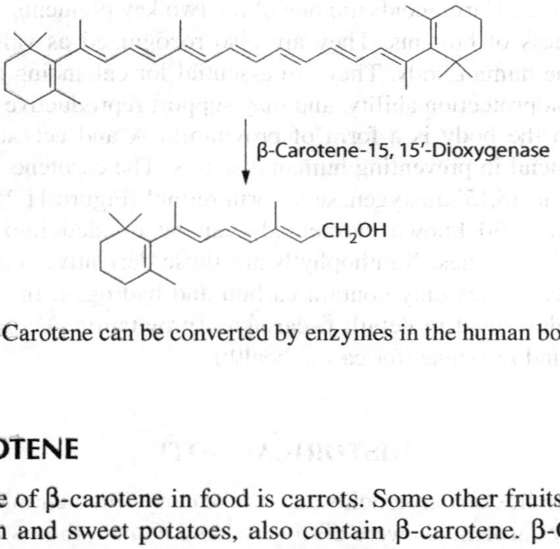

FIGURE 11.2   β-Carotene can be converted by enzymes in the human body to retinol.

11.1   β-CAROTENE

The major source of β-carotene in food is carrots. Some other fruits and vegetables, such as pumpkin and sweet potatoes, also contain β-carotene. β-Carotene is also known as provitamin A and can release vitamin A once hydrolyzed during excursion through the gastrointestinal tract in humans.

11.2   ISOLATION OF β-CAROTENE

β-Carotene can easily degrade or oxidize when exposed to heat, light, or oxygen; therefore the isolation and purification requires care and is often carried out in the absence of light and oxygen, in which case red light, low temperature, and argon gas protection are used. A solvent extraction method can be applied to the isolation and purification of β-carotene using common organic solvents (e.g., hexane, ethanol, acetone, methanol, or tetrahydrofuran) and chromatography over alumina or silica gel.

Supercritical fluid extraction (SFE) methods have replaced liquid solvents: supercritical CO_2 fluid is used as the exaction agent. Compared to solvent extraction, SFE is more attractive, capable of providing mild extraction conditions combined with low energy requirements for solvent recovery. The high selectivity of the extraction process and the reduced potential for oxidation of the extracted materials makes the SFE technique highly suitable for the isolation of many of these natural carotenoid pigments. As an example, pure β-carotene is obtained from tomato paste residue and from microalgae using SFE. However, to maximize the separation efficiency and minimize the degradation and loss, the conditions of pressure, temperature, SFE rate, and the extraction time must be carefully controlled. Specifically, the optimal temperature range for isolation is 40°C to 65°C. High pressure about 200 to 500 bar pressure is required to condense the supercritical CO_2 fluid as the flow rate is maintained around 4 kg/h for 2 hours. Further, ethanol is added as the cosolvent, increasing the yield by 10% to 15%.

Importantly, under conditions of 448 bar and 40°C, the SFE method allows for the efficient separation of the Z/E isomers of β-carotene. The solubility of the 9Z isomer is found to be 4 times higher than that of all the E isomers. Routine analytical analysis to detect β-carotene can be achieved using high-performance liquid chromatography (HPLC) with ultraviolet-visible (UV-Vis) photodiode array (PDA) detection. Best conditions require a short analysis time to minimize the isomerization and decomposition of these sensitive carotenoids.

Reverse phase C_{30}-HPLC with a PDA detector, set from 300 to 700 nm, can be used to separate the carotenoid isomers. For example, the carotenoids in mangoes (*Mangifera indica* L.) are resolved and characterized quantitatively shown in Figure 11.3. The numbered peaks are identified as follows: (1) all-*trans* β-carotene;

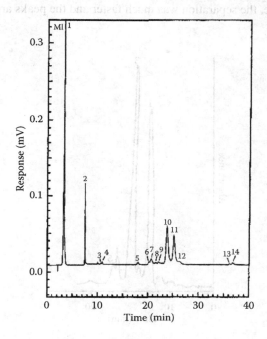

FIGURE 11.3   HPLC separation of carotenoids found in mangoes.

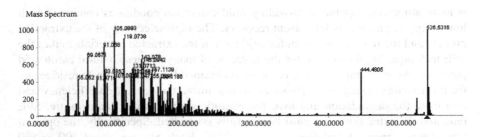

FIGURE 11.4 The mass spectrum of carotene, indicating the MH+ ion and subsequent fragment peaks.

(2) Sudan (used as an internal standard); (3) all-*trans* β-cryptoxanthin; (4) all-*trans* β-crytoxanthin; (5) all-*trans*-β-zeaxanthin (see Section 11.3); (6–9) luteoxanthin isomers; (10) all-*trans*-violaxanthin; (11) 9-*cis*-violaxanthin; (12) 13-*cis*-violaxanthin; (13) *cis*-neoxanthin; and (14) all-*trans*-neoxanthin. The identification of each carotenoid can be confirmed by mass spectrometry (MS). As an example, the MS spectrum of β-carotene shown in Figure 11.4 reveals the molecular ion fragment, MH+, and subsequent fragmentation to smaller ions, 14 mass units (CH2) apart.

β-Carotene has been isolated from other sources, including carrots and algae. In the HPLC chromatograms shown in Figures 11.5 and 11.6, the *cis/trans* isomers of β-carotene have been isolated using a C-30 bonded reverse phase stationary phase. In the former case, the separation was much faster and the peaks are sharper.

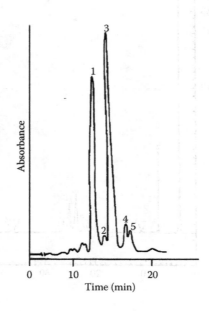

FIGURE 11.5 HPLC separation of carotenoids in carrots.

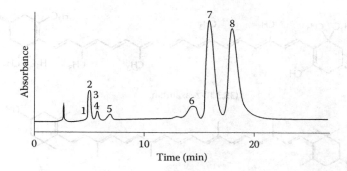

FIGURE 11.6 HPLC separation of carotenoids from an algal source.

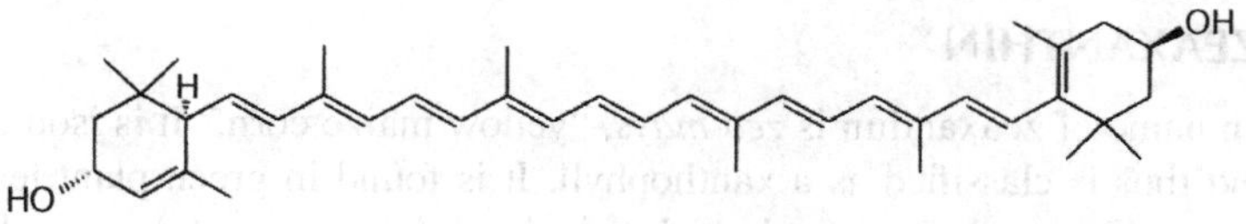

FIGURE 11.7 Chemical structure of lutein.

11.3 LUTEIN

The name *lutein* comes from the Latin *luteus*, meaning "yellow," whereas in Chinese it takes the meaning of "yellow color of leaf." Lutein is synthesized by plants and found in reasonable quantities in green leafy vegetables such as spinach and kale. Lutein is also responsible for the yellow color of egg yolk, chicken skin, and fat.

The long aliphatic carbon chain gives lutein (Figure 11.7) the properties of an antioxidant and as a blue light absorber, thus appearing as a yellow compound—properties that are important for the protection of the human retina. Lutein has a molecular formula, $C_{40}H_{56}O_2$, and has a linear aliphatic carbon chain. It is formed from eight isoprene units (tetraterpenoid) and two hydroxyl functional groups at the two ends.

11.4 ISOLATION OF LUTEIN

Lutein is generally extracted from plants with organic solvents. As an example, from marigolds dichloromethane is used. The base material for extraction is often the saponified oleoresin from the flowers, from which it is possible to obtain a raw crystalline product, enriched in the carotenoids. Using solvent mixtures such as hexane and dichloromethane containing 0.10% N,N-diisoppropylethylamine (DIPEA), crystallization of lutein is achieved at –70°C and crystalline lutein product is obtained of 98% purity. However, if required it can be further purified by HPLC on a silica-based nitrile-bonded column as the stationary phase, eluting with a mixture containing hexane (75%) and dichloromethane (25%) with methanol (0.25%) and DIPEA (0.1%) as the mobile phase.

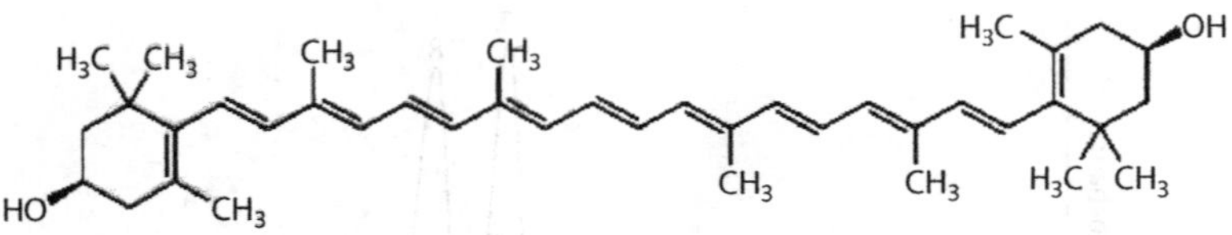

FIGURE 11.8   Chemical structure of zeaxanthin.

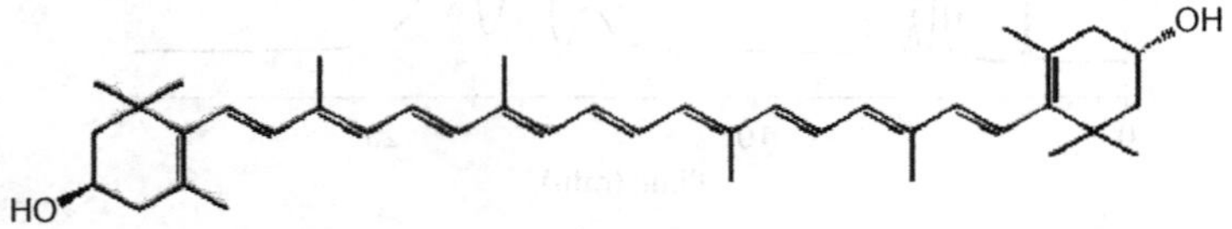

11.5   ZEAXANTHIN

The Latin name of zeaxanthin is *zea mays*, "yellow maize corn." It is isomeric with lutein, and thus is classified as a xanthophyll. It is found in green plant leaves and eggs. Zeaxanthin is also a nonphotochemical quenching agent, accumulating in human eyes at the macula lutea, protecting the eyes from radical damages in a manner similar to lutein. The stereoisomers of zeaxanthin in nature are shown in Figure 11.8.

11.6   ISOLATION OF ZEAXANTHIN

Supercritical CO_2 is used in the isolation of zeaxanthin from red paprika together with other minor carotenoids. The separation is controlled by maintaining the temperature at 60°C. First, the paprika sample is packed into an extraction column, flushed with CO_2 gas to remove O_2. Then, the liquid CO_2 (60°C) is streamed in, and a cosolvent such as ethanol or ethyl ether is added to enhance the extraction. The CO_2 is evaporated off, and "mists" of extracted compounds are deposited in the collector, which can be further separated via centrifugation and deposited on the glass walls.

Analysis of purity can be achieved by HPLC coupled to a UV detector. If further saponification is needed, the material is dissolved in a mixture of methanol and NaOH to generate free zeaxanthin chemical constituents.

11.7   LYCOPENE

The Latin name of lycopene is *lycopersicum*: tomato species. It is a bright, red-colored carotene compound, which widely exists in red fruits and vegetables (e.g., tomatoes, red carrots, red bell peppers, and watermelons).

The chemical structure of lycopene is similar to β-carotene. Both compounds are formed entirely from hydrogen and 40 carbon atoms possess conjugated carbon–carbon double bonds. However, lycopene is a linear (acyclic) carotene and there are two additional carbon–carbon double bonds at both ends of the molecule (Figure 11.9). There are several common geometric isomers of lycopene (Figure 11.10); in total, there are 72 possible isomers.

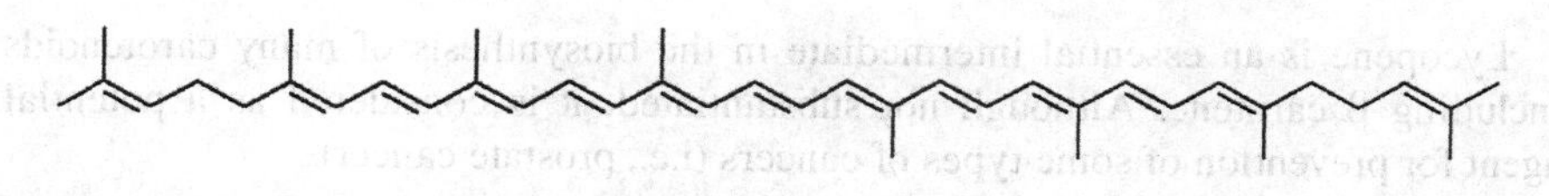

FIGURE 11.9 Chemical structure of lycopene.

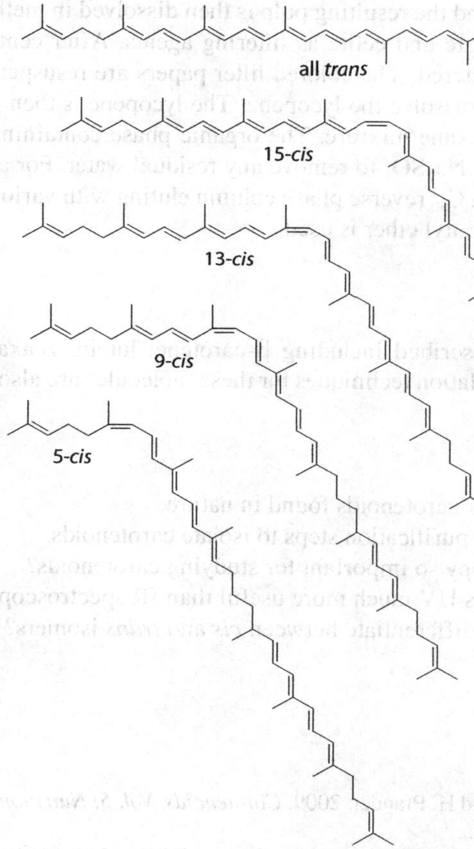

all *trans*

15-*cis*

13-*cis*

9-*cis*

5-*cis*

FIGURE 11.10 Chemical structures representing the various geometrical isomers of lycopene.

The UV-Vis spectrum of lycopene exhibits an absorption maxima (in petroleum ether) of λ_{max} 444, 470, and 500 nm, in comparison to the absorption maxima of β-carotene (λ_{max} at 425 and 450 nm). The reason for the observed shift is the increase in number of unsaturated double bonds in the structure of lycopene compared to β-carotene, which contributes to a shift due to lengthening of the conjugated system. Lycopene shifts its absorption from blue to green color in visible light and appears as a strong red color in contrast to β-carotene.

Lycopene is an essential intermediate in the biosynthesis of many carotenoids including β-carotene. Although not substantiated, it is considered as a potential agent for prevention of some types of cancers (i.e., prostate cancer).

11.8   ISOLATION OF LYCOPENE

Typically, lycopene can be isolated from the juice of tomatoes. First, tomato pulp is obtained by centrifugation and the resulting pulp is then dissolved in methanol, with additions of calcium carbonate and celite as filtering agents. After centrifugation, the colored supernatant is filtered. The colored filter papers are resuspended in the acetone/hexane solvent to redissolve the lycopene. The lycopene is then partitioned using an aqueous acetone/hexane mixture. The organic phase containing lycopene can be dried with anhydrous Na_2SO_4 to remove any residual water. For analysis, an HPLC system consisting of a C_{30} reverse phase column eluting with various ratios of methanol in methyl tertiary butyl ether is used.

SUMMARY

Selected carotenoids are described including β-carotene, lutein, zeaxanthin, and lycopene. Separation and isolation techniques for these molecules are also discussed.

QUESTIONS

1. Give examples of other carotenoids found in nature.
2. Describe some typical purification steps to isolate carotenoids.
3. Why is UV spectroscopy so important for studying carotenoids?
4. For carotenoids, why is UV much more useful than IR spectroscopy?
5. Can UV be helpful to differentiate between *cis* and *trans* isomers?

FURTHER READING

CAROTENOIDS

G. Britton, S. Liaaen-Jensen, and H. Pfander. 2009. *Carotenoids. Vol. 5: Nutrition and Health.* Birkhauser Verlag GmbH.

N. I. Krinsky and E. J. Johnson. 2005. Carotenoid actions and their relation to health and disease. *Mol Aspects Med* 26(6):459–516.

J. T. Landrum, ed. 2009. *Carotenoids: Physical, Chemical, and Biological Functions and Properties.* CRC Press.

Y. Li, A. S. Fabiano-Tixier, V. Tomao, G. Cravotto, and F. Chemat. 2013. Green ultrasound-assisted extraction of carotenoids based on the bio-refinery concept using sunflower oil as an alternative solvent. *Ultrason Sonochem* 20(1):12–18.

A. Z. Mercadante, A. Steck, and H. Pfander. 1999. Carotenoids from guava (*Psidium guajava* l.): Isolation and structure elucidation. *J Agric Food Chem* 47(1):145–151.

D. Ren and S. Zhang. 2008. Separation and identification of the yellow carotenoids in *Potamogeton crispus* L. *Food Chem* 106(1):410–414.

T. L. Sourkes. 2009. The discovery and early history of carotene. *Bull Hist Chem* 34(1):32–38.

Beta-Carotene

I. Gamlieli-Bonshtein, E. Korin, and S. Cohen. 2002. Selective separation of *cis-trans* geometrical isomers of beta-carotene via CO_2 supercritical fluid extraction. *Biotechnol Bioeng* 80(2):169–174.

A. Z. Mercadante, D. B. Rodriguez-Amaya, and G. Britton. 1997. HPLC and mass spectrometric analysis of carotenoids from mango. *J Agric Food Chem* 45(1):120–123.

O. Montero, M. D. Macias-Sanchez, C. M. Lama, L. M. Lubian, C. Mantell, M. Rodriguez, and E. M. de la Ossa. 2005. Supercritical CO_2 extraction of beta-carotene from a marine strain of the cyanobacterium *Synechococcus* species. *J Agric Food Chem* 53 (25):9701–9707.

A. N. Mustapa, Z. A. Manan, C. Y. Mohd Azizi, W. B. Setianto, and A. K. Mohd Omar. 2011. Extraction of β-carotenes from palm oil mesocarp using sub-critical R134a. *Food Chem* 125(1):262–267.

S. Strohschein, M. Pursch, H. Händel, and K. Albert. 1997. Structure elucidation of β-carotene isomers by HPLC-NMR coupling using a C30 bonded phase. *Fresenius' J Anal Chem* 357(5):498–502.

Lutein

F. Khachik. 1999. Process for extraction and purification of lutein, zeaxanthin and rare carotenoids from marigold flowers and plants. Google Patents.

R. L. Roberts, J. Green, and B. Lewis. 2009. Lutein and zeaxanthin in eye and skin health. *Clin Dermatol* 27(2):195–201.

Lycopene

N. Aghel, Z. Ramezani, and S. Amirfakhrian. 2011. Isolation and quantification of lycopene from tomato cultivated in Dezfoul, Iran. *Jundishapur J Nat Pharm Prod* 6(1):9–15.

T. Baysal, S. Ersus, and D. A. J. Starmans. 2000. Supercritical CO2 extraction of β-carotene and lycopene from tomato paste waste. *J Agric Food Chem* 48(11):5507–5511.

12 Selected Vitamins

Vitamins represent a series of organic compounds, which are very crucial to people's health. Some of them can be synthesized by the host organism, while many of them must be obtained from food and are considered essential vitamins. Based on the differences in biological and chemical activities, vitamins are classified into several groups. Each group normally contains a set of compounds and possesses a corresponding function in the body.

The name vitamin was coined as a term derived from "vitamine," a compound word from "vital" and "amine." The classification of vitamin is based on biological and chemical activity but not structure. There are mainly 13 types of vitamins (Vit): A, B_1, B_2, B_3, B_5, B_6, B_7, B_9, B_{12}, C, D, E, and K. In this chapter, five vitamins are discussed: Vit C, Vit D, Vit E, Vit B_6, and Vit B_{12}.

12.1 VITAMIN C

Vit C, also named ascorbic acid, is an essential nutrient for human health. Vit C includes the L-enantiomer of ascorbic acid and its oxidized forms, which have activity in animals, including ascorbic acid and its salts, and some oxidized forms of the molecule. For example, dehydroascorbic acid. D-ascorbate has equal antioxidant power, but is not found in nature, and has no physiological significance, and far less vitamin activity.

Ascorbic acid is a primary metabolite that is directly involved in normal growth, development, and reproduction, and also plays a role as a redox cofactor and catalyst in a broad array of biochemical reactions and processes. Vit C is present in vegetables; fruits; and animal organs such as liver, kidney, and brain.

Vit C can be absorbed in the body by active transportation to all types of cells with the help of several biological transporters as well as simple diffusion. Ascorbate and ascorbic acid are both natural forms in the body, and both of them can interconvert depending on the pH of the environment. Vit C is a cofactor involved in at least eight enzymatic reactions. Furthermore, due to its antioxidant properties, Vit C has been widely used as a food additive, to prevent oxidation, and Vit C can also act as a reducing agent, donating electrons to various enzymatic and nonenzymatic reactions.

12.2 STRUCTURAL FEATURES OF VITAMIN C

The structure of Vit C, with elaboration of the two stereochemical centers, is shown in Figure 12.1. In the ^{1}H spectrum (300 MHz) of Vit C taken in D_2O, there are three peaks corresponding to three types of protons attached to their respective

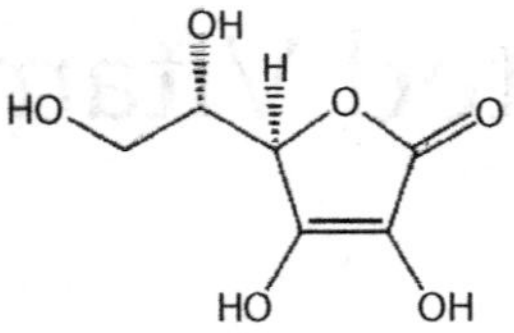

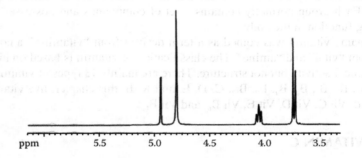

Vitamin C (Ascorbic Acid)

FIGURE 12.1 The structure of vitamin C (ascorbic acid). (R. B. Rucker, J. Zempleni, J. Suttie, and D. McCormick, eds., 2007, *Handbook of Vitamins*, 4th ed., Taylor & Francis.)

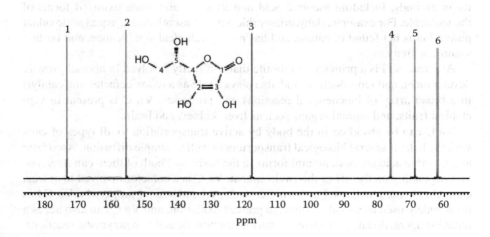

FIGURE 12.2 ^{1}H NMR spectrum (300 MHz) in D$_2$O of vitamin C.

FIGURE 12.3 ^{13}C NMR spectrum of vitamin C with carbon assignments.

carbon atoms. Upon integrating the peak ratios, these signals represent four protons (Figure 12.2).

The ^{13}C NMR spectrum shown in Figure 12.3 indicates the presence of six carbon atoms. In the mass spectrum of Vit C (Figure 12.4), there is a strong molecular ion indicating a m/z 178 mu, with subsequent loss of OH and C$_2$H$_4$O$_2$ fragment ions consistent with the proposed structure.

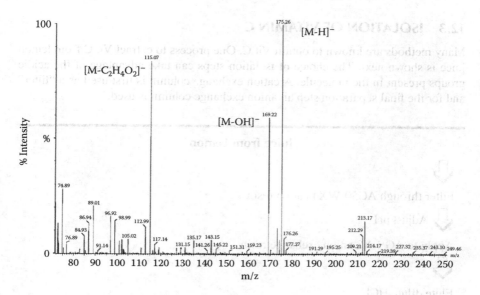

FIGURE 12.4 The mass spectrum of vitamin C.

HISTORICAL NOTE

A published work in 1753 suggested that citrus fruits (limes) contained certain compounds that could treat scurvy. Scurvy was endemic between the 17th and 19th centuries because of insufficient intake of fruits and vegetables. Today, we know that Vit C, also known as ascorbic acid, has the ability to cure and prevent scurvy.

The Vit C discovery began in the late 16th century when French explorers were saved from effects of scurvy by drinking a tea made from the arbor tree during long sea voyages. Later, it was noted that lemon juice can prevent people from getting scurvy and by 1734, it was concluded that people who did not eat fresh vegetables and greens would get the disease, and all seamen were thus provided with citrus fruits. British explorer Cook also supplied his men with limes during their long voyages in the late 18th century. These observations led to important breakthroughs in the understanding of scurvy by conducting experiments on guinea pigs and became one of the first examples of the use of animal models to study nutritional diseases. The first isolation of "ascorbic acid" was achieved in 1937 by Svirbely and Szent-Györgyi, who went on to win the Nobel Prize in Medicine. The first synthesis of Vit C was achieved by Haworth and Hirst, also resulting in a Nobel Prize in Chemistry in 1937. Mass production of Vit C by Hoffmann–La Roche came 20 years later.

12.3 ISOLATION OF VITAMIN C

Many methods are known to obtain Vit C. One process to extract Vit C from lemon juice is shown next. The choice of isolation steps can take advantage of the acidic groups present in the molecule. A cation exchange column is first used as a "filter" and for the final separation step an anion exchange column is used.

Juice from Lemon

Filter through AG50-WX4 cation resin

 Adjust pH

Apply to AG 1-X8 anion exchange resin

Elute dilute HCl

12.4 VITAMIN D

Vit D (Figure 12.5) is not an essential dietary vitamin, as it is synthesized in the body when exposed to sunlight. Vit D was discovered in an effort to understand the cause of rickets, especially in children. Rickets is characterized by impeded growth; and soft, weak, deformity of the bones leading to bow legs as a result of calcium/phosphorus deficiency, as well as a lack of Vit D. Today, Vit D is added to staple foods, such as milk, to avoid this disease due to any nutrient deficiency. Furthermore, the combination of Vit D together with calcium is generally believed to be important for bone health.

The metabolism of Vit D in the body is shown in Figure 12.6. In the liver, Vit D_3 is converted to 25-hydroxyvitamin D_3 (25(OH)D_3). Vit D_2 is converted in the liver to 25-hydroxyergocalciferol, also known as 25-hydroxyvitamin D_2, (25(OH)D_2), shown

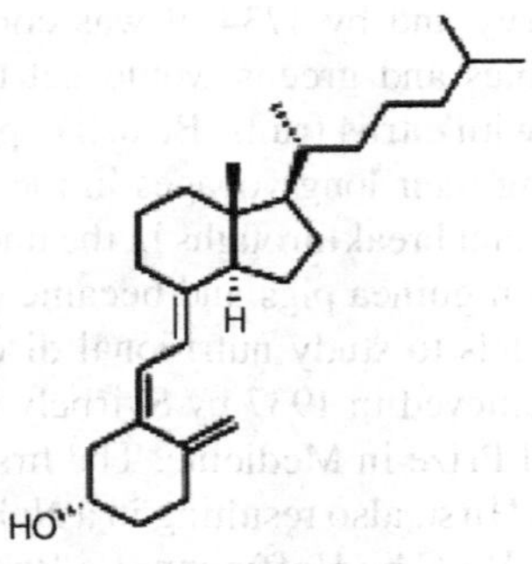

FIGURE 12.5 Chemical structure of vitamin D3.

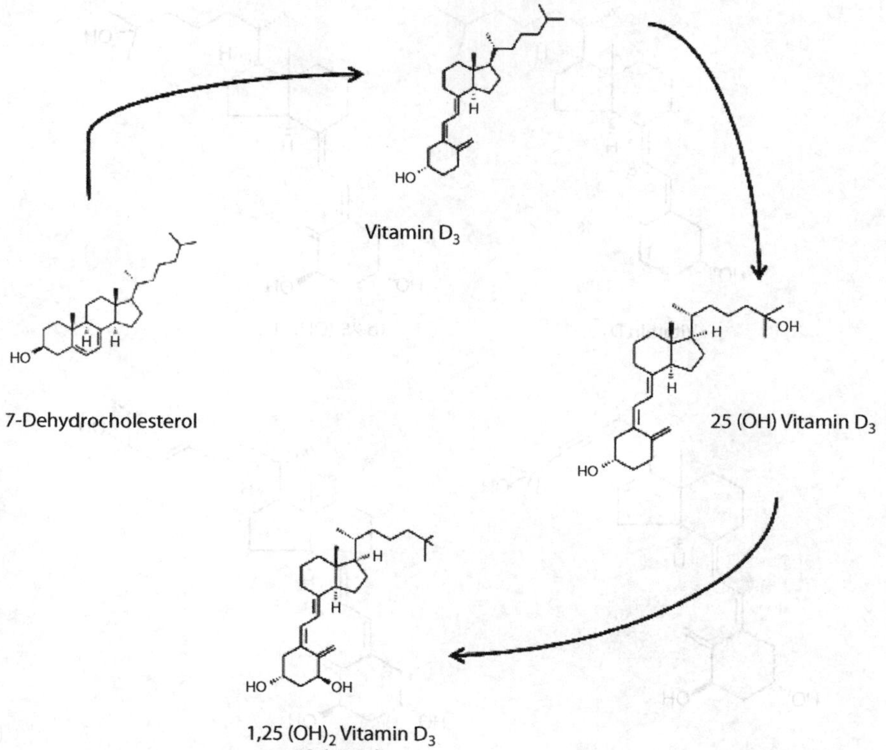

FIGURE 12.6   Vitamin D conversion to key metabolites.

in Figure 12.7. These are the two specific Vit D metabolites that are measured in serum to determine a person's Vit D status.

12.5   VITAMIN E

Vit E represents eight lipid-soluble compounds, including four tocopherols and four tocotrienols, marked by α-, β-, γ-, and δ- separately, and are present in many oils, fruits, vegetables, and foods. Of all the forms of these vitamins, γ-tocopherol is the most common one, while the α-tocopherol is the most biologically active form of Vit E. Based on current research, the symptoms of Vit E deficiency is dependent on α-tocopherol.

Because Vit E cannot be accumulated in the liver, the excretion and metabolism pathways are very important, thus Vit E is an important secondary metabolite and acts as an antioxidant.

Vit E is a very important nutrient to promote efficient health. Generally, Vit E deficiency may result in cardiovascular disease and end-stage renal failure. Supplementation with Vit E alleviates these issues. Currently, Vit E can be found in many foods, vegetables, and fruits.

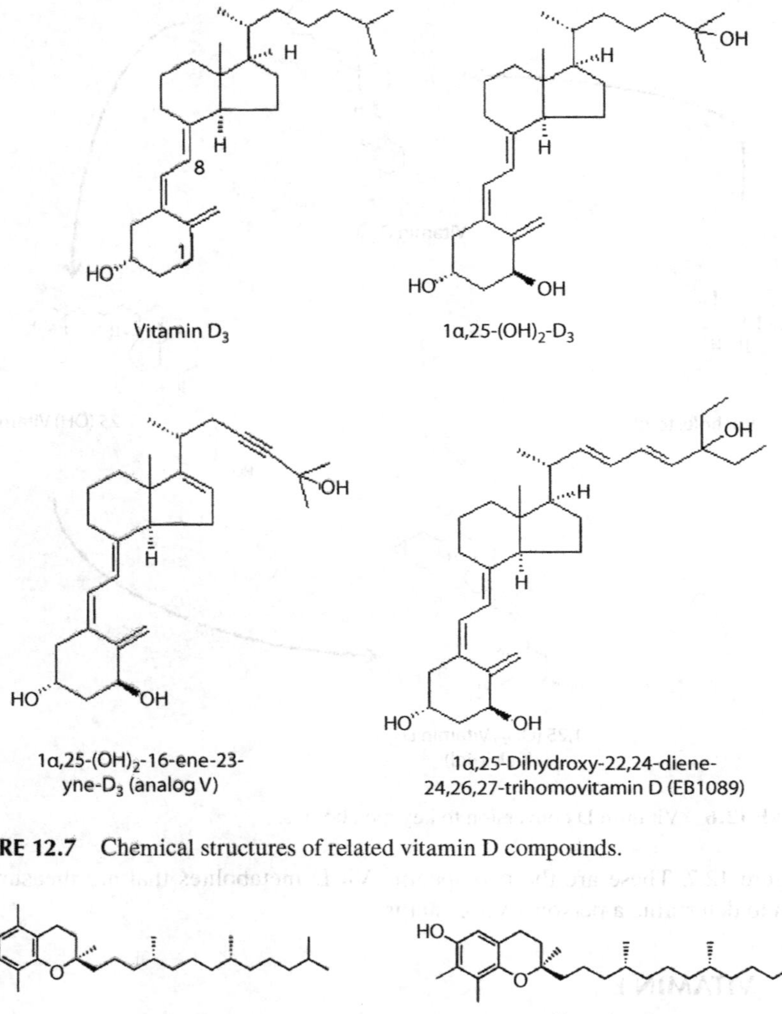

FIGURE 12.7 Chemical structures of related vitamin D compounds.

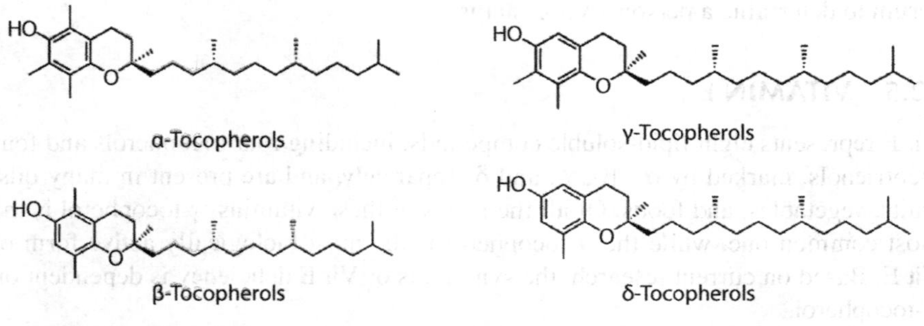

Vitamin D₃

1α,25-(OH)₂-D₃

1α,25-(OH)₂-16-ene-23-yne-D₃ (analog V)

1α,25-Dihydroxy-22,24-diene-24,26,27-trihomovitamin D (EB1089)

α-Tocopherols

γ-Tocopherols

β-Tocopherols

δ-Tocopherols

FIGURE 12.8 Structures of vitamin E-related compounds.

The structure determination of α-tocopherol was completed in 1938, verifying the empirical formula of Vit E as $C_{29}H_{50}O_2$ (Figure 12.8). This is confirmed by the mass spectral data for Vit E and is shown in Figure 12.9.

Since vegetables and fruits contain a relatively low content of Vit E, many methods have been developed to extract Vit E, including liquid extraction, supercritical

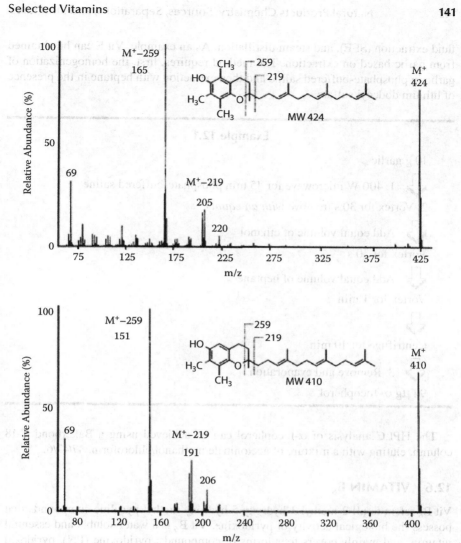

FIGURE 12.9 The mass spectra of vitamin E compounds showing typical fragmentation.

HISTORICAL NOTE

Vit E was first described in 1922. It was isolated from wheat germ and named α-tocopherol. Subsequently, β-tocopherol and γ-tocopherol were separated from vegetable oils. Vit E was first used on preemies (babies born prematurely) who suffered from growth failure, and after supplementation most infants recovered and were able to resume normal growth. In adults, Vit E is a required nutrient with antioxidant benefits and one role may be to prevent oxidative damage caused during lipid peroxidation.

fluid extraction (SFE), and steam distillation. As an example, Vit E can be obtained from garlic based on extraction. The method requires, first, the homogenization of garlic in phosphate-buffered saline and then extraction with heptane in the presence of lithium dodecyl sulfate.

Example 12.1

10 g garlic

⇩   1. 400 W microwave for 45 min phosphate-buffered saline

2. Vortex for 30 s (remove *with an equal*)

⇩   Add equal volume of ethanol

Vortex for 30 s

⇩   Add equal volume of heptane

Vortex for 1 min

⇩

Centrifuge for 10 min

⇩   3. Remove and evaporation

94 μg α-Tocopherol

The HPLC analysis of α-tocopherol can be achieved using a Bakerbond C-18 column, eluting with a mixture of acetonitrile/methanol/chloroform, 47/47/6.

12.6   VITAMIN B$_6$

Vit B$_6$ refers to all 2-methyl-3-hydroxy-5-hydroxy methyl pyridine compounds that possess the biological activity of pyridoxine. Vit B$_6$ is a water-soluble and essential vitamin, and mainly covers four forms of compounds: pyridoxine (PN), pyridoxal (PA), pyridoxamine (PM), and pyridoxal (PL). They can all be phosphorylated by pyridoxal kinase to their 5′-phosphates, known as PNP, PLP, and PMP, respectively. PLP is the metabolically active form, and PNP and PMP can be oxidized to PLP by PM (PN) 5′-phosphate oxidase. The structures of pyridoxine and its related compounds are shown in Figure 12.10.

Pyridoxal phosphate (PA) is biologically an important cofactor in many reactions of amino acid metabolism, including transamination, deamination, and decarboxylation. PLP, in its metabolically active form, can regulate macronutrient metabolism, as well as the synthesis and function of the neurotransmitter, histamine, and hemoglobin, and even gene expression, such as governing the release of glucose from glycogen. Other processes include R-group interconversions; racemization; and alpha-, beta-, and gamma-elimination. In general, PLP enzymes act through Schiff base intermediates with the formation of a resonance stabilized carbanion.

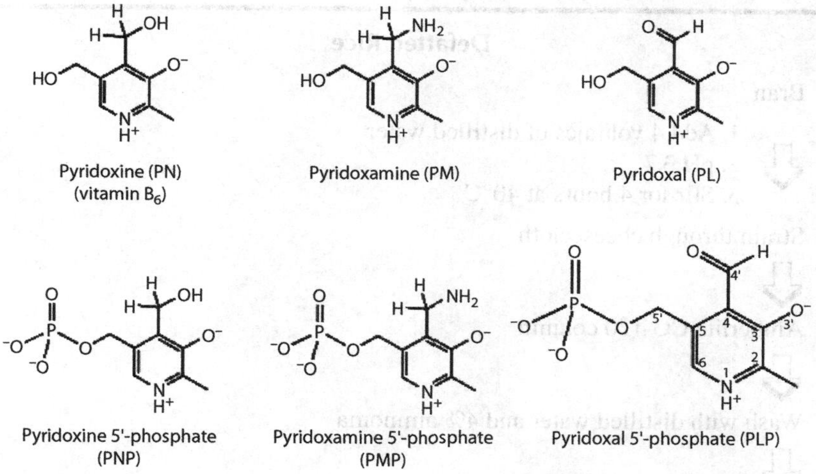

FIGURE 12.10 The structure of pyridoxine and related compounds.

Vit B_6 can offer benefit in the transfer of signals from one nerve cell to another with the participation of neurotransmitters, and be helpful in regulating the body's clock and improving the immune system. The importance of PLP in the body can be recognized by symptoms due to its deficiency. Generally, there is a decrease in appetite and growth rate; with resulting dermatologic symptoms, muscle weakness, dental and oral health issues, and hepatic steatosis. In term of vascular problems, deficiencies cause arteriosclerosis, anemia, paralysis, convulsions, peripheral neuropathy, and malformations. Vit B_6 deficiency will cause confusion, irritability, depression, and mouth and tongue sores in our daily life. Conversely, an excess of Vit B_6 can lead to difficulty in coordinating movement, numbness, and sensory changes.

Since PLP play a major role in amino acid metabolism, it is widely dispersed throughout the plant and animal kingdoms. Common natural sources include rice, fish, poultry, meat, potatoes, fruits (bananas) and vegetables (carrots), and other food sources such as avocado, legumes, and whole grains. Various methods have been used to isolate Vit B_6 from these sources.

HISTORICAL NOTE

Vit B_6 is a secondary metabolite. In 1934, Hungarian physician Paul György noted that it cured skin disease in rats and 4 years later, Vit B_6 was isolated from rice bran. The structure was elucidated by Harris and Folkers in 1939. Subsequently, it was demonstrated that Vit B_6 mainly has two forms: pyridoxal and pyridoxamine. All the forms of Vit B_6 are precursors of PLP, which is now understood to play an important role as a cofactor of many essential enzymes in the human body.

Defatted Rice

Bran

1. Add 4 volumes of distilled water
2. pH 3.7
3. Stir for 4 hours at 40°C

Strain through cheesecloth

Amberlite CG-120 column

Wash with distilled water and 4% ammonia

Obtain the effluent

Brownish syrup

1. Dissolve in water
2. Remove the insoluble substance (5′-O(b-D-glucopyranosyl) pyridoxine)

Bananas

1. Trituration
2. Add buffer and activated charcoal
3. Stir for 0.5 hour and filter

Filtrate

Add ammonium sulfate and mixture of PEG/PVP

Shaking then standing

Collect the organic phase

The various spectral data for Vit B$_6$ are shown in Figures 12.11 and 12.12.

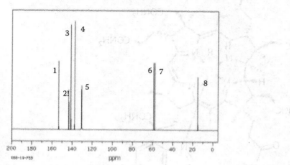

FIGURE 12.11 ¹H NMR spectrum of vitamin B₆. (R. B. Rucker, J. Zempleni, J. Suttie, and D. McCormick, eds., 2007, *Handbook of Vitamins*, 4th ed., Taylor & Francis.)

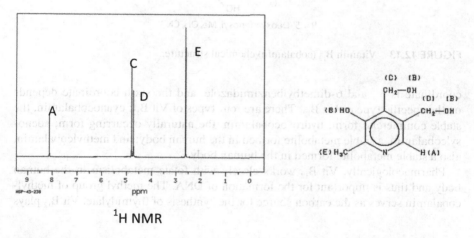

FIGURE 12.12 ¹³C NMR spectrum of vitamin B₆. (J. L. Svirbely and A. Szent-Gyorgyi, 1932, *Biochem J* 26(3):865–870.)

12.7 VITAMIN B₁₂

The Vit B₁₂ (cobalamin) chemical structure is shown in Figure 12.13. It is a fascinating molecule, challenging the ingenuity of the natural products chemist. Over many years of research this elegant structure has generated significant research studies on the chemistry of pyrroles, porphyrins, and corrins, and their respective biosynthetic pathways.

One of the unusual features of the molecule is the carbon–metal bond. Vit B₁₂ is composed of a cobalt atom and a corrin ring, consisting of four pyrrole rings, which gives a porphyrin-like structure. The rings contribute to four coordinates, the fifth

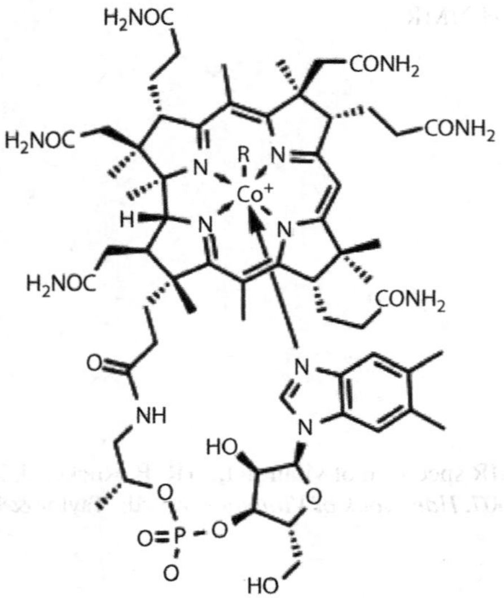

R = 5'-Deoxyadenosyl, Me, OH, CN

FIGURE 12.13 Vitamin B_{12} (cobalamin) chemical structure.

coordinate is 5-, and 6-dimethylbenzimidazole, and the sixth coordinate depends on the specific type of Vit B_{12}. There are four types of Vit B_{12}: cyanocobalamin, the stable commercial form; hydroxocobalamin, the naturally occurring form; adenosylcobalamin, a stable metabolite formed in the human body; and methylcobalamin, also a stable metabolite formed in the human body.

Pharmacologically, Vit B_{12} works closely with folate metabolism in the human body and thus is important for the formation of DNA. The methyl group of methylcobalamin serves as the carbon source for the synthesis of thymidylate. Vit B_{12} plays

HISTORICAL NOTE

The discovery of Vit B_{12} occurred in 1855 out of medical necessity to seek a cure for a mysterious and fatal disease. By 1926, a group of physicians made the remarkable finding that feeding cooked liver to patients suffering with anemia resulted in a cure. In 1928, a liver extract was prepared and this was shown to be much more potent than the liver itself. These findings subsequently led to Whipple, Minot, and Murphy sharing the 1934 Nobel Prize in Physiology for their findings on Vit B_{12}.

Much more study on the structure determination was required. Finally, in 1956, the structure of Vit B_{12} was determined by Hodgkin and her team based on X-ray crystallographic data, for which she was awarded the Nobel Prize in Chemistry.

a key regulatory role in the normal working function of the brain and the nervous system. In fact, high Vit B_{12} levels in elderly individuals may protect against brain atrophy or shrinkage associated with Alzheimer's disease and impaired cognitive function.

Vit B_{12} plays a role in carbohydrate metabolism. Also, it has a role in the synthesis of red blood cells. In the absence of Vit B_{12}, there is no maturation, leading to abnormally large red blood cells.

12.7.1 Structure Elucidation of Vitamin B_{12}

The elucidation of the entire structure of Vit B_{12} can be described in stages of identifying several of the compound's building blocks and piecing them together. This painstaking work was completed without the benefit of today's modern spectroscopic tools. First, the D-1 aminopropan-2-ol structure was found by observing a phosphate associated with a fluorescent substance that gave a ninhydrin reaction under acidic or basic hydrolytic conditions.

The identity of a 5,6-dimethylbenziminazole was revealed after chromatographic and spectroscopic studies of a hydrolyzed fragment. The sugar derivative was isolated as a pentose (through consumption of one mole of periodate). The coordination of the N atom of the 5,6-dimethylbenziminazole moiety to the cobalt metal was shown by spectroscopic studies. The UV absorption maxima at 360 mu, 520 mu, and 550 mu are characteristic of a corrinoid unit, whereas the absorption at 289 mu corresponds to the 5,6 dimethylbenzimidazole substituent.

Researchers showed the complex macro ring structure to be a corrin, not porphyrin, by having a reduced ring structure and one of the meso bridges replaced by a direct linkage. Eventually, the complete structure was confirmed by X-ray crystallography in the 1950s.

The mass spectrum of Vit B_{12} displays a molecular ion $(M+H)^+$, m/z 1355.54. In the NMR spectrum, some assignments were made from 1-dimensional ^1H NMR: methyl (Pr3), and methane (C19H and R1H) multiplicity. Using selective decoupling techniques with comparison to small compounds (axial nucleotide, adenosyl ligand) and spectra of cobalamins with various beta axial ligands and protonation on the benzimidazole-base comparisons, the structure was elucidated.

12.7.2 Commercial Production

Chemical synthesis requires more than 70 steps. Thus, the process is tedious, expensive, and not amenable to large-scale production. Fortunately, fermentation processes using bacteria produce cobalamin derivatives by similar pathways to a chemical synthesis. The production is improved with additions of cobalt ions, glycine, threonine, and choline into the fermentation medium. Upon completion of the fermentation step, the filtered broth can be extracted using an organic solvent such as dichloromethane.

SUMMARY

Several vitamins are described to indicate the variety of the chemistry. They were Vit C, Vit D, Vit E, Vit B_6, and Vit B_{12}.

QUESTIONS

1. List the other recognized vitamins.
2. Provide an example of a structure for each one.
3. What are their important roles/functions in the body?
4. What other atoms might replace Co in Vit B_{12}?
5. How would you determine which atom is present if any?

FURTHER READING

G. F. M. Ball. 2004. *Vitamins: Their Role in the Human Body*. John Wiley & Sons.
R. R. Eitenmiller, W. O. Landen, and L. Ye. 2007. *Vitamin Analysis for the Health and Food Sciences*. 2nd ed. Taylor & Francis.
R. B. Rucker, J. Zempleni, J. Suttie, and D. McCormick, eds. 2007. *Handbook of Vitamins*. 4th ed. Taylor & Francis.

VITAMIN C

A. Albertino, A. Barge, G. Cravotto, L. Genzini, R. Gobetto, and M. Vincenti. 2009. Natural origin of ascorbic acid: Validation by ^{13}C NMR and IRMS. *Food Chemistry* 112 (3):715–720.
I. B. Chatterjee. 1973. Evolution and the biosynthesis of ascorbic acid. *Science* 182(4118): 1271–1272.
M. B. Davies, J. Austin, and D.A. Partridge. 1991. *Vitamin C: Its Chemistry and Biochemistry, (RSC Paperbacks Series)*: Royal Society of Chemistry.
A. Garrido Frenich, M. E. Hernandez Torres, A. Belmonte Vega, J. L. Martinez Vidal, and P. Plaza Bolanos. 2005. Determination of ascorbic acid and carotenoids in food commodities by liquid chromatography with mass spectrometry detection. *J Agri Food Chem* 53(19):7371–7376.
J. Hvoslef. 1968. The crystal structure of L-ascorbic acid, "vitamin C": The x-ray analysis. *Acta Crystallogr B* 24(1):23–35.
J. P. Scannell, H. A. Ax, and W. Morris. 1974. A modern approach to the isolation and characterization of L-ascorbic acid (vitamin C). *J Agri Food Chem* 22(3):538–539.
J. L. Svirbely and A. Szent-Gyorgyi. 1932. The chemical nature of vitamin C. *Biochem J* 26(3):865–870.
J. G. Wang, Y. N. Wang, and L. Y. Ma. 2010. The isolation and determination of vitamin C from orange peel. *Sci Technol Innovation Herald* 14:136–137.
J. X. Wilson. 2005. Regulation of vitamin C transport. *Annu Rev Nutr* 25:105–125.

VITAMIN E

M. Birringer, P. Pfluger, D. Kluth, N. Landes, and R. Brigelius-Flohe. 2002. Identities and differences in the metabolism of tocotrienols and tocopherols in HepG2 cells. *J Nutr* 132(10):3113–3118.
O. H. Emerson, G. A. Emerson, A. Mohammad, and H. M. Evans. 1937. The chemistry of vitamin E: Tocopherols from various sources. *J Bio Chem* 122:99–107.
E. Fernholz. 1938. On the Constitution of α-Tocopherol. *J Am Chem Soc* 60(3):700–705.
P. Karrer, H. Fritzsche, B. H. Ringier, and H. Salomon. 1938. Synthese des α-tocopherols. *Helvetica Chimica Acta* 21(1):820–825.
A. E. Lindberg. 2010. *Vitamin E: Nutrition, Side Effects and Supplements*. Nova Science.

M. N. Malik, M. D. Fenko, A. M. Shiekh, and H. M. Wisniewski. 1997. Isolation of α-tocopherol (vitamin E) from garlic. *J Agri Food Chem* 45(3):817–819.

VITAMIN B$_6$

E. E. Snell. 1944. The vitamin activities of "pyridoxal" and "pyridoxamine." *J Bio Chem* 154(1):313–314.

E. T. Stiller, J. C. Keresztesy, and J. R. Stevens. 1939. The structure of vitamin B$_6$. I. *J Am Chem Soc* 61(5):1237–1242.

Xiexiu-Juan and Zhang Zhen-Xin. 2011. Determination of trace vitamin B$_6$ by fluorimetry after aqueous two phase extraction. *Acta Nutrimenta Sinica* (1):99–100.

K. Yasumoto, H. Tsuji, K. Iwami, and H. Mitsuda. 1977. Isolation from rice bran of a bound form of vitamin B$_6$ and its identification as 5′-O-(b-D-glucopyranosyl) pyridoxine. *Agri Bio Chem* 41(6):1061–1067.

VITAMIN B$_{12}$

R. Banerjee. 1999. *Chemistry and Biochemistry of B12.* Wiley.

A. Bax, L. G. Marzilli, and M. F. Summers. 1987. New insights into the solution behavior of cobalamins. Studies of the base-off form of coenzyme B$_{12}$ using modern two-dimensional NMR methods. *J Am Chem Soc* 109(2):566–574.

R. Bonnett. 1963. The chemistry of the vitamin B$_{12}$ group. *Chem Rev* 63 (6):573–605.

X. Luo, B. Chen, L. Ding, F. Tang, and S. Yao. 2006. HPLC-ESI-MS analysis of vitamin B$_{12}$ in food products and in multivitamins-multimineral tablets. *Analytica Chimica Acta* 562(2):185–189.

A. Makarov and J. Szpunar. 1999. Species-selective determination of cobalamin analogues by reversed-phase HPLC with ICP-MS detection. *J Analy Atomic Spectromet* 14(9):1323–1327.

Y. Piao, M. Yamashita, N. Kawaraichi, R. Asegawa, H. Ono, and Y. Murooka. 2004. Production of vitamin B$_{12}$ in genetically engineered *Propionibacterium freudenreichii*. *J Biosci Bioeng* 98(3):167–173.

Z. Zhicong, Z. Heng, and G. Juefen. 2011. Extraction and purification of vitamin B$_{12}$ in fermentation by macroreticular LX-20. *Ion Exchange Adsorp* 27(4):368–374.

Section IV

Nature's Pleasures and Dangers

13 Natural Products in Food, Spices, and Beverages

13.1 SWEETENERS

Natural sweeteners are very popular as an alternative sweetening agent to sugarcane or fructose corn syrup. Sugarcane is the granulated regular table sugar, obtained from a tropical grass, by pressing to extract the juice, then boiling, cooling, and allowing it to crystallize into granules. The chemistry of alternative sweeteners is extensive: from simple sugars to other chemicals. Examples include honey, brown rice syrup, maple syrup, molasses, xylitol, aspartame, sucralose, stevia, and agave.

13.2 XYLITOL

Xylitol $(CHOH)_3(CH_2OH)_2$ (Figure 13.1) is used as a diabetic sweetener and is seen as beneficial for dental health by reducing caries to a third in regular use. It is found in the fibers of fruits and vegetables. On a commercial scale, xylitol is obtained as a hydrolysis product of hemicellulose from corncobs.

13.3 ASPARTAME

In contrast, aspartame (Figure 13.2) is an artificial sweetener. It is a methyl ester of the aspartic acid/phenylalanine dipeptide.

13.4 SUCRALOSE

Sucralose (Figure 13.3) is also an artificial sweetener. However, its structure is very similar to sucrose (see Figure 5.4) with the addition of two substituted chlorine atoms. It is approximately 1000 times sweeter than sucrose.

13.5 STEVIA

Stevia (Figure 13.4) is derived from a perennial shrub and the leaves are 30 times sweeter than sugar. Since the active ingredient possesses zero calories, this plant may be useful for people with diabetes, hypoglycemia, or suffering from candida fungal infections as a low-carbohydrate, low-sugar food alternative.

The genus stevia contains about 240 species of herbs and shrubs in the sunflower family (Asteraceae) and is native to subtropical and tropical regions from western North America to South America. The species *Stevia rebaudiana*, commonly

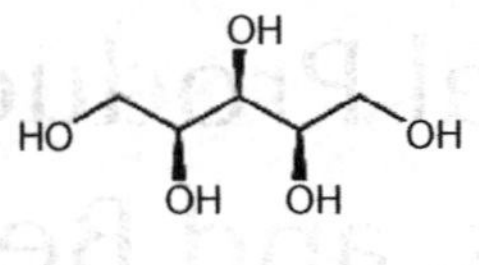

FIGURE 13.1   Chemical structure of xylitol.

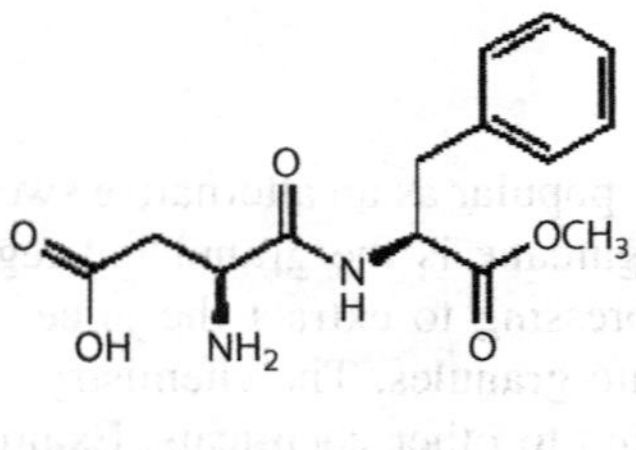

FIGURE 13.2   Structure of the sweetener aspartame.

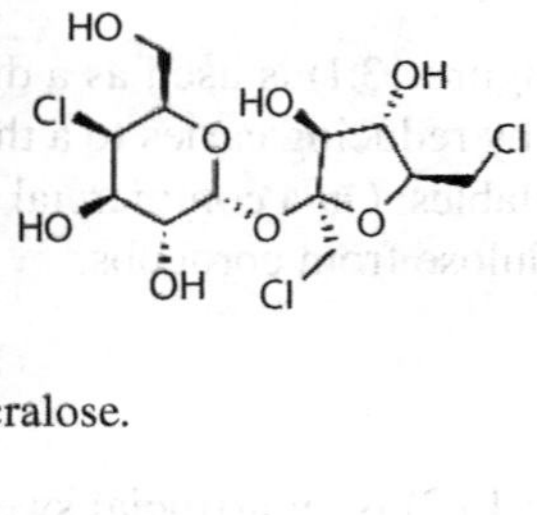

FIGURE 13.3   Structure of sucralose.

known as sweetleaf, is widely grown for its sweet leaves. Stevia's taste has a slower onset and longer duration than that of sugar, although some of its extracts may have a bitter or licorice-like aftertaste at high concentrations.

The steviol glycoside extracts have up to 300 times the sweetness of sugar. The tongue's taste receptors react to the glucose in the glycosides; those with more glucose (rebaudioside) taste sweeter than those with less (stevioside). Some of the tongue's bitter receptors react to the aglycones. Further, in the digestive tract, rebaudiosides are metabolized into stevioside. Then stevioside is broken down further into glucose and steviol. The glucose released in this process is used by bacteria in the colon and is not absorbed into the bloodstream. Steviol cannot be further digested and is passed out from the digestive system in the urine or feces.

13.6   LICORICE

Glycyrrhizin (also known as glycyrrhizic acid or glycyrrhizinic acid) is the main sweet-tasting compound from licorice root and is 30 to 50 times as sweet as sucrose

FIGURE 13.4   *Stevia rebaudiana*. (Photograph courtesy of Steven Foster.)

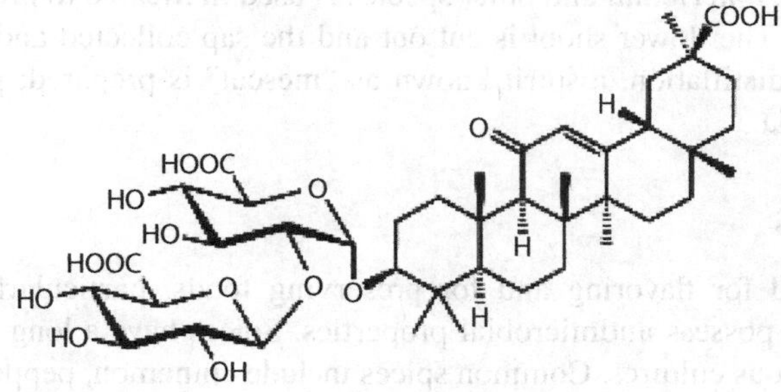

FIGURE 13.5   Chemical structure of glycyrrhizic acid from *Glycyrrhiza glabra*.

(Figure 13.5). Glycyrrhizin is a triterpenoid saponin glycoside. Upon hydrolysis, the glycoside loses its sweet taste and is converted to the aglycone glycyrrhetinic acid.

13.7   AGAVE

Agave nectar is produced from the juice of the core of the agave, a succulent plant native to Mexico. Agave is extracted from the agave cactus plant (Figure 13.6). It

FIGURE 13.6 Agave is extracted from the agave cactus plant and used as a sweetener. (Courtesy of Steven Foster.)

is sweeter than sugar and may be suitable for diabetics. Agave juice is extracted, filtered, heated, and hydrolyzed into agave syrup. It contains 90% fructose and trace amounts of iron, calcium, potassium, and magnesium.

There are four major parts of the agave that are edible: the flowers, the leaves, the stalks or basal rosettes, and the sap. The stalks, especially if roasted, are sweet. The agave nectar, called agave syrup, is a sweetener derived from the sap and used as an alternative to sugar.

The sap of *A. americana* and other species is used in Mexico to produce an alcoholic beverage. The flower shoot is cut out and the sap collected and subsequently fermented. By distillation, a spirit known as "mescal" is prepared; perhaps better known as tequila.

13.8 SPICES

Spices are used for flavoring and for preserving foods, particularly meat products. They also possess antimicrobial properties. Spices have a long history of use throughout various cultures. Common spices include cinnamon, pepper, cloves, nutmeg, cumin, and ginger.

Spices are known as antioxidants, due to the presence of phenolic compounds and in particular the flavonoids. As an example, turmeric (*Curcuma longa*) is a plant of the ginger family. The roots are boiled and ground into the deep orange-yellow powder commonly used as a spice in curry preparations. The active ingredient is curcumin (Figure 13.7), and is now being studied further for its medicinal and health benefits.

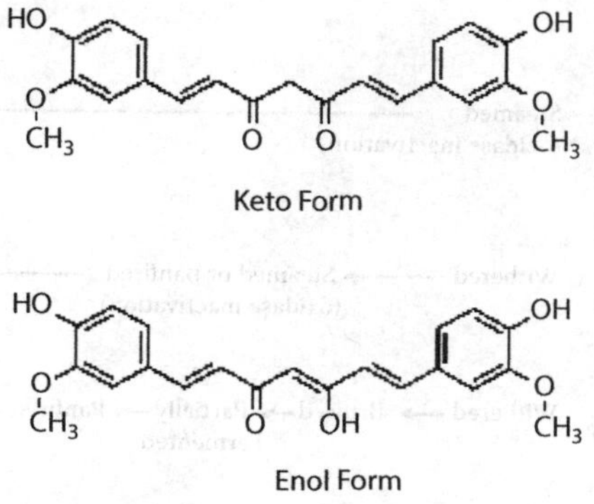

Keto Form

Enol Form

FIGURE 13.7   Structure of curcumin representing keto-enol tautomerism.

13.9   TEA

13.9.1   The Growing and Processing of Tea

Green, black, and white teas are all derived from *Camellia sinensis*, an evergreen shrub of the Theaceae family. Unlike black tea, which is formed by oxidation of green tea after picking, green tea is harvested and usually supplied in its natural state with no further processing. Green tea may be consumed in the form of a brewed beverage. Successful tea cultivation requires moist humid climates typically found on the slopes of Northern India, Sri Lanka, Tibet, and Southern China. Green tea is consumed predominantly in China, Japan, India, and a number of countries in North Africa and the Middle East, whereas black tea is consumed predominantly in Western and some Asian countries.

After the leaves are picked, they undergo one or more of the following processes, which determine whether the final tea product will remain green or white, or transform to oolong, or black teas. Figure 13.8 describes the processing of tea leading to determining the type of tea and the characteristic flavonoid content.

Withering—Fresh, green leaves and buds are softened by withering. The leaves are allowed to air dry in the sun or are placed on racks in a large, heated room. During this stage, the starch in the leaves begins to convert to sugar. For white tea, leaves are withered for 4 to 5 hours; for green, oolong, and black tea, almost twice as long.

Rolling—After withering, leaves are rolled, a process originally performed by hand but now handled by machine, that twists and crushes the leaves, releasing the sap and exposing it to oxygen, which stimulates further oxidation.

Oxidation—For white and green tea, this step is skipped. All black teas are fully oxidized. Oolong tea, however, is only partially oxidized. Rolled

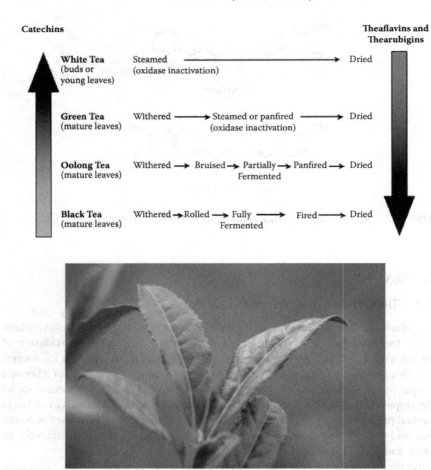

FIGURE 13.8 Processing of tea determines the type of tea and the characteristic flavonoid content. (Photograph courtesy of Steven Foster.)

leaves are placed on trays and left in a cool, damp place for 1 to 3 hours and the leaves turn from green to a copper color.

Drying—After the rolled leaves are oxidized (in the case of black tea), they are dried with hot air to quickly stop the oxidative process or to prevent mold.

13.9.2 TEA CATECHINS

The tea contains beneficial antioxidants, but high-quality green and white teas possess greater concentrations than black tea. Today, scientists believe that catechins are the main active ingredients of green tea. The catechins include the polyphenolic compounds known as epicatechin (EC), epicatechin-3-gallate (ECG), epigallocatechin (EGC), and epigallocatechin-3-gallate (EGCG), all of which may be responsible for the anticarcinogenic and antimutagenic activities of green tea (Figure 13.9). Other

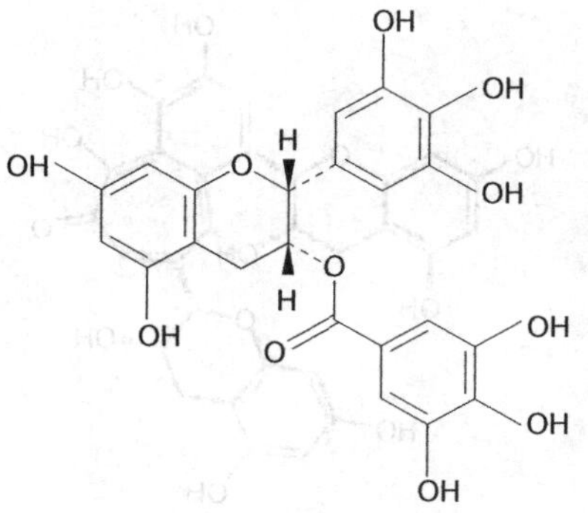

Epigallocatechin gallate (EGCG)
Epigallocatechin (EGC)
Epicatechin (EC)
Epicatechin gallate (ECG)
Catechin (C)
Gallocatechin (GC)

FIGURE 13.9 Chemical structures of the green tea polyphenols, collectively known as the catechins.

HISTORICAL NOTE

First eaten in cheese-like balls until mankind discovered fire and created pots to boil water, tea has now achieved a status as the second most popular drink in the world. No one really knows when the leaves of *Camellia sinensis* were first used to make tea. A legend dating to 3000 BC tells that the mythical Chinese emperor and his followers had stopped to rest under a small tree before continuing with their journey. The emperor was warming a pot of water on a fire when a leaf from the tree fell into the water. The emperor drank it and was said to have immediately recognized the health benefits of the plant. Archeological evidence actually predates this legend and suggests that tea was first consumed during the early Paleolithic period (about 5000 years ago).

Trade between China and the West allowed for the spread of tea. Portugal was the first country to set up trade with Asian countries, but other European countries soon followed. In 1600, Queen Elizabeth granted a charter to what became the British East India Company for the sole purpose of promoting trade with Asia. Tea was first brought to London in 1657 as a medicinal herb and was originally found only in the apothecaries. By the end of the 17th century, both black and green teas were being shipped to England from China in great quantities. The British were passionate about the idea of growing tea in India and by the middle of the 19th century more than a half million acres of tea were planted in India.

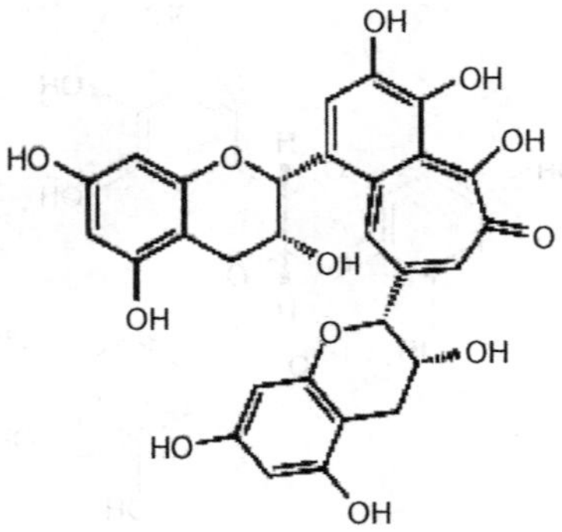

FIGURE 13.10 Chemical structures of the theaflavins found mostly in black tea.

polyphenols in green tea include flavanols and their glycosides, caffeine, small amounts of methylxanthines, and the amino acid theanine.

Polyphenols in green tea have been identified as strong antioxidants and they possess anticarcinogenic properties. Human studies on the pharmacokinetics of these polyphenols suggest that these ingested compounds and their metabolites may play a role in the action against gastrointestinal cancers.

Many *in vitro* and *in vivo* studies demonstrate that the polyphenols from green tea are anticarcinogenic by inducing apoptosis and inhibiting cell growth. Probable action mechanisms include antioxidant and free-radical scavenging activity and stimulation of detoxification systems through selective induction or modification of phase I and II metabolic enzymes.

13.9.3 THEAFLAVINS

Theaflavins (Figure 13.10) are antioxidant polyphenols not found in green tea but present in black tea after the oxidative process. They are formed from the flavan-3-ols (EGCg) in green tea leaves during the enzymatic oxidation of the tea leaves, thus imparting the black tea color and flavor. Theaflavin-3-gallate, theaflavin-3′-gallate, and theaflavin-3-3′digallate are all types of thearubigins, which are reddish in color.

SUMMARY

Natural products are found in foods, spices, and beverages. Natural products are also used as preservatives for foods (spices) and as sweeteners. Examples include the important natural sweeteners: stevia, licorice, and agave, followed by examples of chemicals in spices and the polyphenolics in teas. Teas have had an important role during human civilization for more than 5000 years.

QUESTIONS

1. Provide the isolation steps to isolate the representative sweetener compound from stevia, licorice, and agave.
2. What methods are used to isolate the catechins from tea?
3. How might you choose to separate the individual components?
4. What is keto-enol tautomerism?
5. How might you detect the equilibrium of these two forms?

FURTHER READING

SWEETENERS

L. O'Brien-Nabors. 2011. *Alternative Sweeteners*. 4th ed. Taylor & Francis.

CATECHINS AND GREEN TEA

R. Cooper, D. J. Morre, and D. M. Morre. 2005. Medicinal benefits of green tea. Part I. Review of non-cancer health benefits. *J Alternative and Complementary Medicine* 11:521–528.
R. Cooper, D. J. Morre, and D. M. Morre. 2005. Medicinal benefits of green tea. Part II. Review of anti-cancer properties. *J Alternative and Complementary Medicine* 11:639–652.
I. MacFarlane and A. MacFarlane. 2009. *The Empire of Tea: The Remarkable History of the Plant That Took over the World*. Overlook Press.
L. C. Martin. 2007. *Tea: The Drink That Changed the World*. Tuttle Publishing.

14 Toxins in Nature

One of the biggest "poison" myths is the suggestion that "all-natural" or "organic" substances are safe and safer than the man-made or derived "chemical" counterparts. The fact is some of the most toxic substances known are from nature. This chapter provides examples of toxins drawn from various sources: plants, mushrooms, spiders, marine algae, fish, snails, and frogs. The examples not only represent the diversity of the chemical sources found in nature but how their different chemical structures contribute to severe toxic effects and in most cases can lead to death.

14.1 EXAMPLES OF TOXINS FROM PLANTS

Aristolochic acids—Aristolochic acids (Figure 14.1) are commonly used in Chinese herbal medicine even though they are a family of carcinogenic mutagenic and nephrotoxic compounds found in *Aristolochia* and *Asarum* (wild ginger). Extreme care of preparation and use is required.

Monkshood (*Aconitum napellus*)—The toxin in this plant is known as aconitine (Figure 14.2 and Figure 14.4b). Aconitine and related alkaloids found in the Aconitum species are highly toxic cardiotoxins and neurotoxins. Symptoms include tingling, numbness and paralysis, violent gastrointestinal symptoms, and intense pain. Death is usually due to cardiac arrhythmias. However, there are traditional uses of this plant. For example, the roots of *Aconitum ferox* yield the Nepalese poison called *nabee*. It contains large quantities of the alkaloid pseudaconitine, which is a deadly poison and used by various Asian communities as arrow poisons to hunt ibex and bear. In Traditional Chinese Medicine, aconite roots are used only after processing to reduce the toxic alkaloid content. However, the use of a larger than recommended dose and inadequate processing increases the risk of poisoning. The wild plant (especially the roots) is extremely toxic. Reduction of the toxic alkaloid content is achieved by soaking and boiling to hydrolyze aconite alkaloids into less toxic and nontoxic derivatives.

Water hemlock (*Cicuta* species)—The toxic principle cicutoxin (Figure 14.3 and Figure 14.4a) is known for causing seizures. Fatal cases have been reported as a result of people eating a water hemlock plant they thought was a wild parsnip or other root vegetable. One mouthful can be fatal and in the United States water hemlock is considered one of North America's most toxic plants to humans.

Castor bean (*Ricinus communis*)—Castor beans contain toxic albumins, which are cellular toxins. As little as one seed fully chewed and swallowed can be fatal in a child. The toxins are identified as the Ricin A chain and Ricin B chain and are of similar molecular weight. Thus, the Ricin A chain is

FIGURE 14.1 Chemical structure of aristolochic acid, toxin in *Aristolochia littoralis*. (Photograph courtesy of Steven Foster.)

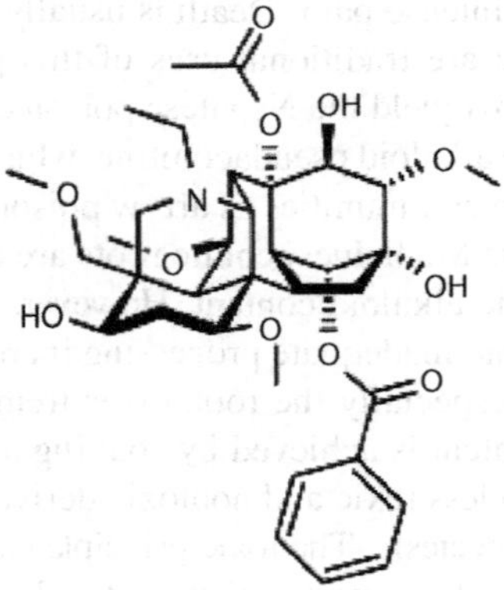

FIGURE 14.2 Chemical structure of the plant toxin aconitine from *Aconitum napellus*.

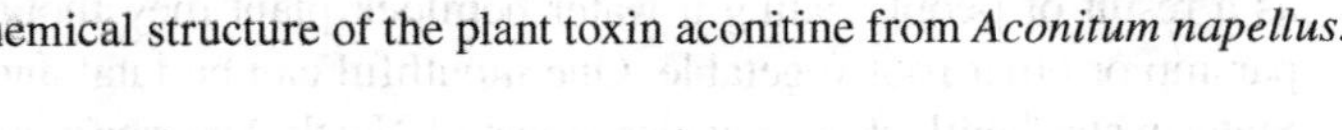

FIGURE 14.3 Chemical structure of cicutoxin from the *Cicuta* plant species.

FIGURE 14.4 Examples of toxic plant species. (a) *Cicuta species*. (b) The plant *Aconitum napellus*. (Courtesy of Steven Foster.) *(Continued.)*

composed of 267 amino acids as a polypeptide arranged into alpha-helices and beta-sheets. The Ricin B chain is composed of 262 amino acids that is able to bind terminal galactose residues on cell surfaces. These toxic albumins can be deliberately harvested from the plant and made into the warfare agent known as ricin.

FIGURE 14.4 (Continued)   Examples of toxic plant species. (c) The plant *Aritsolochia littoralis*. (Courtesy of Steven Foster.)

14.2   TOXIC MUSHROOM: *AMANITA* SP.

The most toxic mushrooms are of the *Amanita* species (*A. verosa, A. verna,* and *A. phalloides*). Just one mushroom cap can be fatal if ingested. The toxins in these mushrooms are identified as cyclic peptides. The structure of α-amanitin is shown in Figure 14.5, and careful isolation from the amanita mushroom is shown as follows:

Isolation

Ground mushroom

    Extract with 1:1 v/v methanol-water

Retain the supernatant

    Column chromatography
Powdered cellulose mixed with acetone
Solvent: 20:6:5:1 parts of methyl ethyl ketone, acetone, water, and butanol

α-Amanitin and β-Amanitin

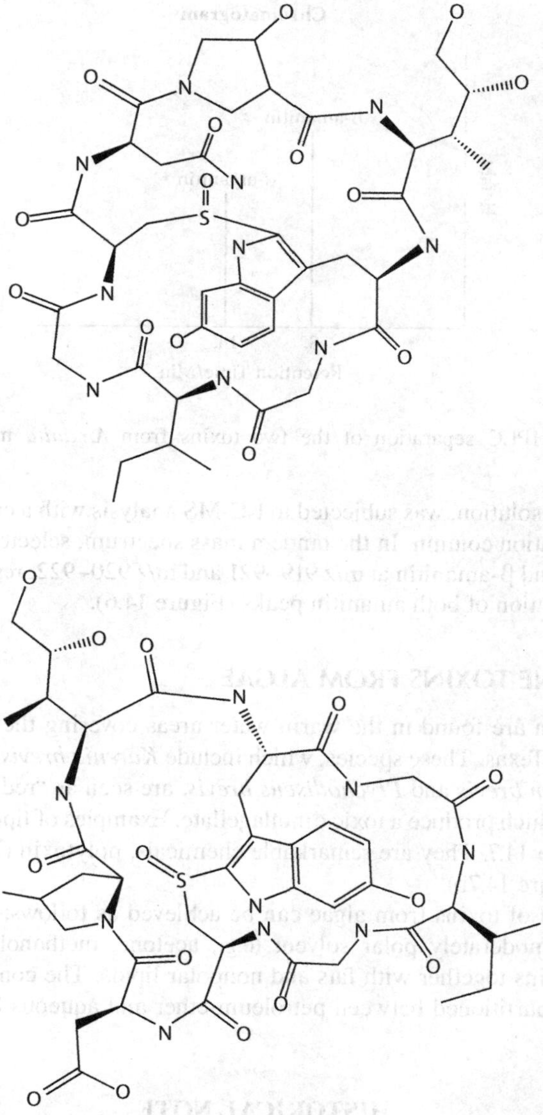

FIGURE 14.5 α- and β-Amanitin.

These compounds are highly toxic and can destroy the liver. Because of the concerns over toxicity, forensic analysis of α-amanitin and β-amanitin in human plasma by liquid chromatography-mass spectrometry has been investigated. Good separation has been achieved using high-performance liquid chromatography–mass spectrometry (LC-MS) and LC–tandem mass spectrometry (LC-MS-MS) analysis. A Discovery DSC 18 column was used, eluting first with water and then increasing amounts of methanol. The subsequent extract, after evaporation and reconstitution

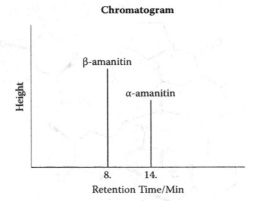

FIGURE 14.6 HPLC separation of the two toxins from *Amanita* mushroom: α- and β-amanitin.

in mobile phase solution, was subjected to LC-MS analysis with a conventional octadecyl LC separation column. In the tandem mass spectrum, selected ion monitoring of α-amanitin and β-amanitin at m/z 919–921 and m/z 920–922, respectively, generates good separation of both amanitin peaks (Figure 14.6).

14.3 MARINE TOXINS FROM ALGAE

Sources of toxin are found in the warm water areas covering the Gulf of Mexico, from Florida to Texas. These species, which include *Karenia brevis,* formerly known as *Gymnodinium brevis* and *Ptychodiscus brevis,* are seen as "red tides" caused by algal blooms, which produce a toxic dinoflagellate. Examples of lipophilic toxins are shown in Figure 14.7. They are remarkable chemicals, polytoxin (Figure 14.7a) and brevitoxin (Figure 14.7b).

The isolation of toxins from algae can be achieved as follows: extraction of the algae using a moderately polar solvent (e.g., acetone, methanol, or acetonitrile) releases the toxins together with fats and nonpolar lipids. The concentrated solvent extract is then partitioned between petroleum ether and aqueous 80% methanol to

HISTORICAL NOTE

Red tide toxins, which are produced from algal blooms, were found along the Pacific Coast of the United States and led to paralytic shellfish poisoning (PSP). An early description was given over 200 years ago by Captain George Vancouver exploring the waters of Vancouver Island and Puget Sound in Washington State. The crewmen became ill after eating mussels. This poisoning resulted in death by paralysis, due to collapse of the respiratory system. However, the complete structure of the PSP toxin was only determined 30 years ago.

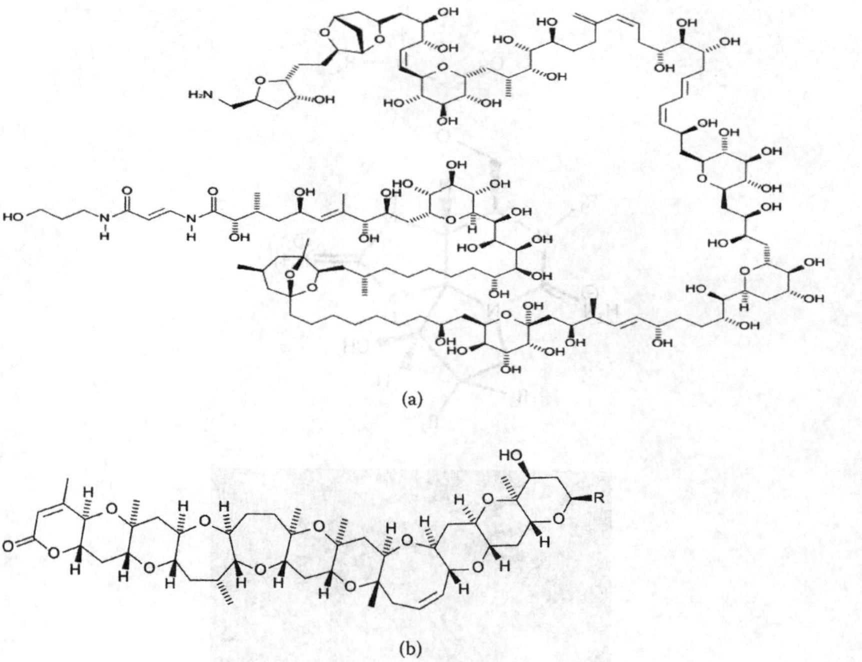

(a)

(b)

FIGURE 14.7 Structures of polytoxin (a) and brevitoxin (b); complex cyclic polyether compounds.

remove the unwanted lipids, and the brevitoxins remaining in aqueous methanol are purified further by chromatography.

The structure of the most potent component of the PSP toxins is called saxitoxin and referred to as STX (Figure 14.8). Saxitoxin is a low molecular weight, water-soluble, nitrogen-containing compound. The key feature is the carbamoyl moiety leading to making saxitoxin a carbamate derivative, which is quite toxic to humans. Various isomers exist (different groups attached to the basic STX molecule) that form a very potent suite of toxins.

14.4 MARINE TOXINS: FISH

Two of the most well-known and most deadly fish poisons are known as ciguatoxin and tetrodotoxin, both of which affect the nervous system. Ciguatoxin (Figure 14.9) is the most potent sodium channel toxin known, and ciguatera poisoning is associated with reef dwelling fish such as barracuda, grouper, and snapper. Tetrodotoxin (Figure 14.10) is associated with the puffer fish/blowfish, known in Japan as "fugu," considered a delicacy but if improperly prepared by not removing the toxic "sac" can lead to death. Chemically, these two toxins are wonderful examples of polyether compounds, requiring extensive and careful spectroscopic and structure determination. In the case of polytoxin, Kishi et al. at Harvard University came up with a remarkable total synthesis to confirm the stereochemistry.

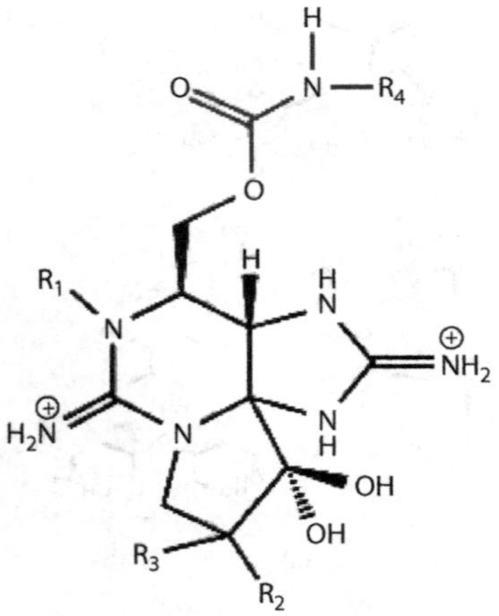

FIGURE 14.8 The family of saxitoxin compounds.

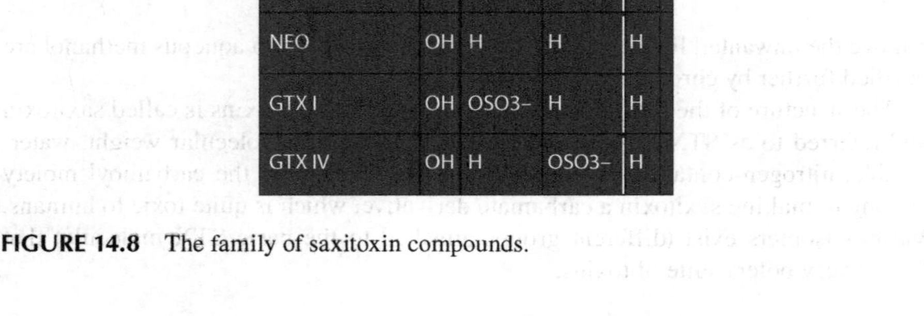

Carbamoyl Saxitoxins	R1	R2	R3	R4
STX	H	H	H	H
GTX II	H	OSO3–	H	H
GTX III	H	H	OSO3–	H
NEO	OH	H	H	H
GTX I	OH	OSO3–	H	H
GTX IV	OH	H	OSO3–	H

FIGURE 14.9 Chemical structure of ciguatoxin (named CTX1B).

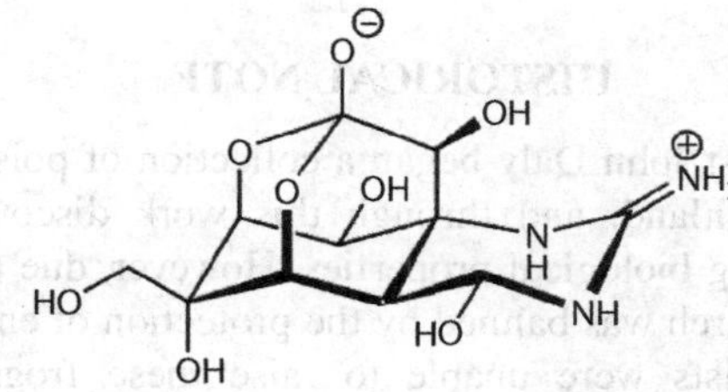

FIGURE 14.10 The chemical structure of the toxin obtained from the blowfish "fugu" named tetrodotoxin.

14.5 SPIDER VENOM

Agelenopsis aperta, commonly known as the desert grass spider, is a species of spider belonging to the family Agelenidae. It is found in dry regions across the southern United States and into Mexico. The venom of the spider is composed of a family of proteins that function as neurotoxins. The biological mechanism of action of the spider toxin is the blockage of calcium channels or acting as potassium channel inhibitory toxins. An example is the compound hanatoxin, a 35-amino acid peptide toxin isolated from the Chilean rose tarantula venom, which inhibits the voltage-gated K-channel by altering the energetics of gating.

14.6 CONUS SNAIL TOXINS

Conus snails are large mollusks and they use a toxic venom to immobilize their prey. The toxin is a mixture of conopeptides and usually consists of between 10 and 30 amino acids. Their mode of action targets different voltage- and ligand-gated ion channels. Several conopeptides have been taken into the clinic and one peptide has received approval for the treatment of pain. The peptides are produced synthetically using solid phase peptide synthesis.

14.7 POISONOUS FROGS

Poisonous substances can be found on the skin of a kind of poison dart frog named *Epipedobates tricolor*. These frogs inhabit the rainforests of Ecuador and the poison is used by Amerindian tribes to make poison arrows. The compound epibatidine (Figure 14.11), a new class of alkaloid, was discovered in 1992, exhibiting a rare example of an alkaloid with a Cl substituent. Biological tests showed it is a nonopioid

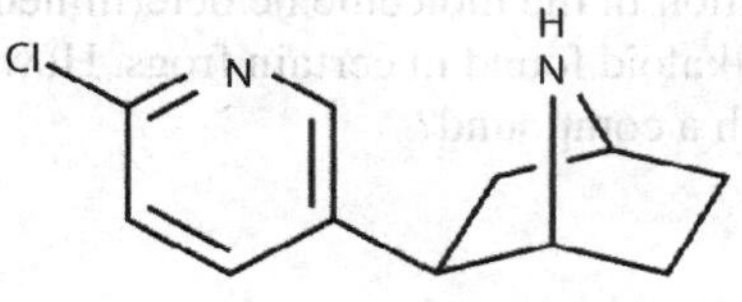

FIGURE 14.11 The chemical structure of the toxin epibatidine (1R,2R,4S)-(+)-6-(6-chloro-3-pyridyl)-7 azabicyclo-[2.2.1]heptane.

HISTORICAL NOTE

In the 1970s, scientist John Daly began a collection of poisonous frogs from the Ecuadorian Highland, and through this work discovered an alkaloid possessing interesting biological properties. However, due to ecological considerations, the research was banned by the protection of endangered species. Furthermore, scientists were unable to raise these frogs with secretion-containing alkaloids in their laboratories.

analgesic compound and hope was given this would be a lead compound for analgesic drug development.

Chemically, epibatidine is a hygroscopic oily substance that can act as a base. The five-membered ring is part of the 7-azabicyclo [2.2.1] heptane structure with exo-oriented chloro and pyridyl substituents.

From X-ray diffraction studies epibatidine is levorotatory and the configuration is established as 1R, 2R, and 4S. The frog skin extracts were further purified on a pre-packed silica gel 60 column (chloroform-methanol-aqueous ammonia and after concentration purified by HPLC using aqueous $(NH_4)_2CO_3$ as a solvent). Biologically, these actives behave as neurotoxins, with high affinity to the acetylcholine receptors: nicotinic and muscarinic acetylcholine receptors. Thus, they have the effect of causing analgesia at low doses, however, reports of paralysis, loss of consciousness, coma, and even death were seen at high doses.

SUMMARY

This final chapter highlights nature's toxic compounds. Examples are shown of spider, conus snail, and frog venoms. Examples of marine toxins from algae include the saxitoxins and brevitoxins. Marine toxins from fish include ciguatoxin and tetrodotoxin.

QUESTIONS

1. How might you isolate red tide toxins from marine sources?
2. What spectroscopic techniques would be useful to help and finally determine the structure of a brevitoxin compound?
3. How might you isolate the saxitoxin compounds?
4. Some saxitoxins contain an SO3 group. How might this group be identified?
5. How might its position in the molecule be determined?
6. Epibatidine is an alkaloid found in certain frogs. How might you undertake the isolation of such a compound?

FURTHER READING

MARINE TOXINS

R. W. Armstrong, J.-M. Beau, S. H. Cheon, W. J. Christ, H. Fujioka, W.-H. Ham, L. D. Hawkins, H. Jin, S. H. Kang, Y. Kishi et al. 1989. Total synthesis of palytoxin carboxylic acid and palytoxin amide. *J Am Chem Soc* 111:7530.

J. K. Cha, W. J. Christ, J. M. Finan, H. Fujioka, Y. Kishi et al. 1982. Stereochemistry of palytoxin. 4. Complete structure. *J Am Chem Soc* 104:7369–7371.

M. Forino, P. Ciminiello P. et al. 2010. Palytoxins: A still haunting Hawaiian curse. *Phytochemistry Reviews* 9:491–500.

R. E. Moore and P. J. Scheuer. 1971. Palytoxin—New marine toxin from a coelenterate. *Science* 172:495–498.

K. C. Nicolau, Z. Yang, G. Shi, J. L. Gunzner, K. A. Agrios, and P. Gärtner. 1998. Total synthesis of brevetoxin A. *Nature* 392:264–269.

E. M. Suh and Y. Kishi. 1994. Synthesis of palytoxin from palytoxin carboxylic acid. *J Am Chem Soc* 116:11205.

MUSHROOMS

D. A. Bushnell, P. Cramer, and R. D. Kornberg. 2002. Structural basis of transcription: Alpha-amanitin-RNA polymerase II cocrystal at 2.8 A resolution. *Proc Natl Acad Sci USA* 99(3):1218–1222.

M. Cochet-Meilhac and P. Chambon. 1974. Animal DNA-dependent RNA polymerases. 11. Mechanism of the inhibition of RNA polymerases B by amatoxins. *Biochim Biophys Acta* 353(2):160–184.

F. Enjalbert, S. Rapior, J. Nouguier-Soule, S. Guillon, N. Amouroux, and C. Cabot. 2002. Treatment of amatoxin poisoning: 20-year retrospective analysis. *J Toxicol Clin Toxicol* 40(6):715–757.

FROGS

X. Rui and B. Donglu. 1999. Chemistry and pharmacology of new potent analgesic epibatidine. *Progress Chem* 11 (03):313.

T. F. Spande, H. M. Garraffo, M. W. Edwards, H. J. C. Yeh, L. Pannell, and J. W. Daly. 1992. Epibatidine: A novel (chloropyridyl)azabicycloheptane with potent analgesic activity from an Ecuadoran poison frog. *J Am Chem Soc* 114(9):3475–3478.

Z.-P. Zhong, L. Zhu, Z.-L. Huang, and C.-J. Cao. 2000. Progress on total synthesis of epibatidine. *Chinese J Synth Chem* 8:486–492.

Glossary (And Some Companies Offering Select Resins)

Pharmacognosy

Spectroscopy
 Nuclear Magnetic Resonance, NMR
 2-Dimensional NMR, 2D-NMR
 Proton NMR, ^{1}H-NMR (^{1}H-NMR)
 Carbon ^{13}NMR, ^{13}C-NMR (^{13}C-NMR)
 Mass Spectrometry, MS,
 X-ray Crystallography, X-ray
 Circular Dichroism, CD
 Ultraviolet Light, UV
 Infra Red Spectroscopy, IR

Detectors
 Photo Diode Array Detector, PDA
 Electrochemical Ionization Detector

Separation Modes
 Gas Chromatography, GC
 Super Critical Fluid Chromatography, SFE
 Flash Chromatography
 Thin Layer Chromatography, TLC
 Capillary Electrophoresis, CE
 Capillary Zone Electrophoresis, CZE
 Ion Chromatography (Dionex)

Column Chromatography Packing Materials
 Cellulose powder
 Alumina gel
 Silica gel
 Polyamide gel
 Charcoal

Ion Exchange Resin Materials
 BioRadAG1x8
 BioRad AG50x8

Liquid Chromatography Approaches
   High Performance Liquid Chromatography, HPLC
   Reversed Phase Chromatography-HPLC Columns
   C2 up to C18, C-30, Nitrile Column

Normal Phase Chromatography
Silica Bonded Phase
CarboWax 20M
Sephadex LH20
Polyamide gel
Counter Current Chromatography, CCC
MCI CHP20P gel (Mitsubishi, Japan)
HP-20
XAD-2, –4, –8,-16 (Rohm and Haas)

Index

9781466567610